T0265190

Cognitive Systems Engineering in Health Care

Cognitive Systems Engineering in Health Care

Edited by
Ann M. Bisantz
Catherine M. Burns
Rollin J. Fairbanks

CRC Press
Taylor & Francis Group
Boca Raton   London   New York

CRC Press is an imprint of the
Taylor & Francis Group, an **informa** business

CRC Press
Taylor & Francis Group
6000 Broken Sound Parkway NW, Suite 300
Boca Raton, FL 33487-2742

Printed on acid-free paper
Version Date: 20141017

International Standard Book Number-13: 978-1-4665-8796-0 (Hardback)

Visit the Taylor & Francis Web site at
http://www.taylorandfrancis.com

and the CRC Press Web site at
http://www.crcpress.com

A.M.B.

To Jack

C.M.B.

To Gary and Jack

R.J.F.

To Mary Linda

Contents

Acknowledgments

The authors acknowledge Robert Wears, MD, PhD, for his insights and inspiration in the science of safety in health care, and Erica Savage for her project management expertise in the preparation of this book.

Editors

Ann M. Bisantz, MS, PhD, is a professor and the chair of the Department of Industrial and Systems Engineering at the University at Buffalo (UB), the State University of New York. Dr. Bisantz holds a doctorate degree in industrial and systems engineering/human machine systems from Georgia Institute of Technology. She has extensive research experience in human–computer interface design and evaluation research for complex systems such as health care. Her research includes developing novel information displays for complex systems, advancing methods in cognitive engineering, and modeling human decision making; she has worked extensively in the domains of health care and defense. She has collaborated with health informatics researchers and clinicians on research regarding health IT usability, workflow impacts, and human factors of electronic health records and has conducted patient safety studies including risk analysis studies and simulation and field studies of emergency department patient-tracking systems (funded by the American Society for Health Care Risk Management, the Emergency Medicine Foundation, and the Agency for Health Care Research on Quality). She is a past recipient of a CAREER award from the National Science Foundation, which has supported research activities in the area of visualization of uncertainty and dynamic decision making, has received funding from NASA and a number of defense organizations, and has been involved with collaborative research with the Center for Multisource Information Fusion at UB. Dr. Bisantz was selected as an attendee (2001) and a speaker (2008) at the National Academy of Engineering's Frontiers in Engineering Symposium (topic: Cognitive Engineering Applications in Health Care) and was recognized with a Young Investigator Award from the UB in 2002. She was appointed ISE department chair in 2012.

Catherine M. Burns, MSc, PhD, is a professor of systems design engineering and the founding director of the Centre for Bioengineering and Biotechnology at the University of Waterloo, Waterloo, Ontario, Canada. She is a world leader in the method of cognitive work analysis with substantial contributions applying and developing the method. This method forms the basis for ecological interface design, the approach in her first book, which has shown remarkable success at improving decision making in complex work situations across a variety of domains. She conducts research in ecological interface design, cognitive work analysis, graphical interface design, and visualization. Her research is applied to health care and health informatics, finance, driver assistance, and consumer products. She has coauthored 7 books and published more than 200 publications in the field. She has received both teaching and research excellence awards at the University of Waterloo. Her research is funded by the Natural Sciences and Engineering Research Council Canada (NSERC), the Canadian Institutes for Health Research (CIHR), and multiple industrial partners, attracting more than $10 million in funding in the past five years.

Rollin J. Fairbanks, MD, MS, FACEP, is the director of the National Center for Human Factors in Healthcare, the director of the Simulation & Training Environment Laboratory (MedStar SiTEL), and an attending emergency physician at the MedStar Washington Hospital Center, which are all part of MedStar Health in Washington, DC. He is an associate professor of emergency medicine at Georgetown University and an adjunct associate professor of industrial and systems engineering at the University at Buffalo. Dr. Fairbanks holds a master's of science degree in industrial systems engineering and human factors engineering from Virginia Tech and a medical degree from the Virginia Commonwealth University's Medical College of Virginia. His work applies safety science to medical systems, with a focus on human factors, system safety engineering, health information technology, and medical device design. His research has been funded by the National Institutes of Health, AHRQ, ONC, and several foundations, and he has published more than 100 papers in the medical and engineering literature.

Contributors

Shilo H. Anders
Cognitive Solutions Division
   Applied Research Associate
Fairborn, OH

Ann M. Bisantz
University at Buffalo
The State University of New York
Buffalo, NY

Jeffrey P. Brown
Cognitive Solutions Division
   Applied Research Associate
Fairborn, OH

Catherine M. Burns
University of Waterloo
Waterloo, ON, Canada

Lindsey N. Clark
National Center for Human Factors
   in Healthcare, MedStar Health
Washington, DC

Beth Crandall
Cognitive Solutions Division
   Applied Research Associate
Fairborn, OH

Rollin J. Fairbanks
National Center for Human Factors
   in Healthcare, MedStar Health
Washington, DC

Anna Grome
Cognitive Solutions Division
   Applied Research Associate
Fairborn, OH

Theresa K. Guarrera
University at Buffalo
The State University of New York
Buffalo, NY

Zachary A. Hettinger
National Center for Human Factors
   in Health Care, MedStar Health
Washington, DC

David T. LaVergne
University at Buffalo
The State University of New York
Buffalo, NY

Nicolette M. McGeorge
University at Buffalo
The State University of New York
Buffalo, NY

Laura Militello
Applied Design Science, A, LLC
Cincinnati, OH

Anne Miller
Center for Research and Innovation
   in Systems Safety, School
   of Nursing & Department
   of Biomedical Informatics,
   Vanderbilt University Medical
   Center
Nashville, TN

Christopher P. Nemeth
Cognitive Solutions Division
   Applied Research Associate
Fairborn, OH

Jeremy C. Pamplin
Army Institute for Surgical
    Research San Antonio Military
    Medical Center
San Antonio, TX

Sarah Henrickson Parker
National Center for Human Factors
    in Health Care, MedStar Health
Washington, DC

Avi Parush
Department of Psychology
Carleton University
Ottawa, ON, Canada

Priyadarshini R. Pennathur
Department of Mechanical and
    Industrial Engineering
University of Iowa
Iowa City, IA

Shawna J. Perry
Virginia Commonwealth University
    Medical
School of Medicine
Richmond, VA

C. Adam Probst
Baylor Scott & White Health
Dallas, TX

Yan Xiao
Baylor Scott & White Health
Dallas, TX

1

Cognitive Engineering for Better Health Care Systems

Ann M. Bisantz, Rollin J. Fairbanks, and Catherine M. Burns

CONTENTS

Introduction

The discipline of cognitive engineering began more than 30 years ago in response to newly recognized challenges faced by people working in and controlling complex, high-technology systems. These challenges resulted in part from a fundamental shift of work away from physical labor to sophisticated cognitive tasks such as sense making, planning, diagnosing, and decision making. System designers and researchers recognized that the cognitive work demands imposed by these systems—characterized by complex, interconnected, and dynamic components; increasing levels of automation; and an ever-wider reach—could lead to errors with potentially disastrous outcomes. In fact, high-profile accidents in domains of nuclear power (Three Mile Island; Chernobyl), aviation (Air France Flight 296), and military systems (USS Vincennes; USS Stark) both demonstrated the need for, and led to advanced research in, a new framework of cognitively informed system design—cognitive engineering.

More recently, the call for systems approaches to improve patient safety (Kohn et al. 1999), combined with increased deployment of information technology throughout health care, has led to recognition of the role that cognitive engineering can play in ensuring safe and effective healthcare. This book, through a set of case examples and research studies, provides evidence of that impact.

Brief Overview of Cognitive Engineering

Goals

The goal of cognitive engineering (also called cognitive systems engineering) is to support human performance in the context of complex, technological systems characterized by complexity, uncertainty, time pressure, and risk (Woods and Roth 1988). Methods in cognitive engineering are used to understand and represent two important and complementary perspectives of such human–technology systems: challenges and constraints on performance stemming from complexities in the work environment itself, and the knowledge and strategies used by experienced practitioners in those environments to perform successfully (Vicente 1999; Crandall et al. 2006; Bisantz and Roth 2008). Outputs from cognitive engineering analyses are used to support system understanding and design.

Methodological Approaches

Methods in cognitive engineering encompass both data-gathering techniques and analytical approaches and models. Across methodological approaches, cognitive engineering analyses take a consistent position that data gathering must focus on an understanding derived from domain experts and based on the study of activities and situations in actual work environments. Typically, this is accomplished through observation of actual work situations; interviews or focus groups with domain experts; review of technologies, artifacts, and documents used in the work setting; and in some cases, participation of subject matter experts on the analysis team. For instance, the study described in Chapter 4 involved subject matter experts (emergency physicians) on the analysis and design team, and a number of chapters explicitly describe the use of interviews (Chapters 3, 4, 8, and 9) and observational methods (Chapters 3, 7–9). Chapters 6, 8, and 9 focus heavily on the role that devices, technologies, and other "low-tech" artifacts have in shaping work performance.

There are a number of structured methods used in cognitive engineering, including cognitive work analysis (CWA) (Vicente 1999), cognitive task analysis (CTA) (Crandall et al. 2006), goal-directed task analysis (Endsley et al. 2003), applied cognitive work analysis (ACWA) (Elm et al. 2003), applied cognitive task analysis (Militello and Hutton 1998), and work-centered system design (Eggleston 2003). Despite variations in approach, all provide methods for framing data collection and analysis while balancing (to a greater or lesser extent) the dual goals of capturing domain practitioners' knowledge, skills, and strategies as well as work domain complexities and tasks. For example, CTA often focuses heavily on knowledge and expertise of expert domain practitioners. Structured interview techniques, such as the multipass critical incident interviewing technique, are used to elicit characteristics of expert decision making and performance in challenging situations. CWA focuses more explicitly on complexities in the work domain by including five phases, which capture (1) purposes, processes, resources, and constraints associated with the work domain; (2) information processing stages associated with typical system control tasks; (3) strategies used by human or automated agents to accomplish those tasks; (4) constraints and requirements stemming from the socio-technical organization; and (5) knowledge and skills required for successful system control. Chapter 3 applies this approach to a team-intensive hospital environment. ACWA shares a focus on system purposes and processes with CWA but calls out more explicitly the links among system aspects, key decisions, information requirements, and ultimate design products. It is beyond the scope of this book to provide a detailed overview of these methodological approaches; however, readers can refer to Bisantz and Roth (2008) for a comprehensive review of methods, Vicente (1999) and Crandall et al. (2006) for in-depth treatments of the CWA and CTA methodologies, and Bisantz and Burns (2008) for a review of CWA applications.

Outcomes

Cognitive engineering analyses have a wide range of design-oriented outcomes, many of which are illustrated in this volume. Typical outputs include innovative information display concepts (Chapters 4 and 5), concepts for decision-support systems (Chapter 2), methods or technologies to support team communication and coordination (Chapters 3, 5–8), strategies addressing adaptive allocation of activities between people and automated systems, design guidance for devices or information systems (Chapters 2 and 9), recommendations regarding individual or team training (Chapter 3), and methods to help manage workload (Chapter 9).

Chapter Contributions: Applications and Themes

The chapters contained in this book highlight current cognitive engineering–oriented research, analyses, and applications in a variety of settings, including cardiac surgery, obstetrics, and emergency medicine.

In addition to covering a range of health care settings, the chapters can be viewed within four broad themes as follows:

1. There is a consistent focus, throughout many of the chapters, on the impact that cognitive engineering analyses can have in supporting communication and coordination within health care teams—in

labor and delivery (Chapter 3), emergency departments (Chapter 4), surgery (Chapters 5 and 8), intensive care (Chapter 7), and throughout a patient's hospital stay (Chapter 6).

2. Several chapters demonstrate the use of cognitive engineering methods to inform the design of information technology. Chapter 2 describes the role of cognitive engineering analyses in designing clinical decision support systems and health information technology more generally. Chapter 4 presents novel display concepts for an emergency department information system intended to support tracking and management of patients seeking emergency care. The chapter demonstrates both the cognitive engineering–based analyses and the iterative, user-centered design method that led to the designs. Chapter 5 presents the design for a team-oriented display to support cardiac surgery, which emphasizes both extensive data collection with a surgical team and a systematic method for grouping information so that it could best support team performance.

3. A number of chapters demonstrate the systematic adaptation and application of specific cognitive engineering methods in the medical domain. For instance, Chapter 3 describes the adaptation and use of multiple phases of CWA to document and understand team interactions within a hospital labor and delivery unit. Chapter 4 describes the application of work domain analysis (one phase of CWA) to understand the complexities of a hospital emergency department. Chapter 6 describes a new model that documents traces of information that emerge during the interaction of people, work, and technology, and illustrates the model using a patient treatment case. Chapter 9 demonstrates the combination of cognitive engineering–informed analyses with macroergonomic approaches to address real-world problems within a large hospital system.

4. Several chapters provide examples of how in-depth cognitive engineering analyses can lead to demonstrated improvements in health care environments. Chapter 7 presents a detailed case study of a burn intensive care unit, which will be used to design artifacts supportive of decision making and communication. Chapter 8 describes the development of a number of interventions to improve surgical team performance and communication during cardiac surgery. Chapter 9 describes the impact of a combined cognitive engineering/macroergonomic approach in solving challenging problems related to nursing workload, cardiac surgery, electronic health record (EHR) documentation, and drug infusion across a large health care system.

Conclusions

This book will provide the reader with an understanding of the value of the cognitive systems engineering approach in health care, and, through a series of sample studies, will help the reader understand how these approaches might be applied in the health care domain.

In reading this book, we hope that the reader will notice some common themes in the application of cognitive engineering to health care. First, the domain of health care is complex, and understanding it builds a deep respect for the expertise of the practitioners who work in this domain. The understanding of disease and best practices in care are constantly evolving. Cognitive engineering seeks to build solutions in this transitioning environment by searching for abstractions that hold constant while technologies, practices, and understanding continue to evolve.

Further, to understand these environments, researchers often adapt and combine multiple methods to reach the best understanding that they can of the environment. The methods of cognitive engineering act as lenses on the world, and changing lenses can add valuable new perspectives. This adaptation, in itself, is an example of creative and innovative practice in cognitive engineering.

Finally, in the context of a book, we were able to allow the authors the space to express their ideas more fully than in a journal paper. This allows the methods to be fully discussed and the transition to design to be shown more clearly. The deeper discussions allowed here are important as it allows the small gaps to be filled and a richer understanding of practice to be developed. There is a need for further sharing of deeply described studies and data so that the field can continue to develop.

As with all work, this book is merely a snapshot in time. We expect the application of cognitive engineering in health care to continue to grow and evolve with new applications, new challenges, and new approaches.

References

Bisantz, A. M., and Burns, C. M. (2008). *Applications of Cognitive Work Analysis*. Boca Raton, FL: CRC Press.

Bisantz, A. M., and Roth, E. M. (2008). Analysis of cognitive work. In D. A. Boehm-Davis (Ed.), *Reviews of Human Factors and Ergonomics* (Vol. 3, pp. 1–43). Santa Monica, CA: Human Factors and Ergonomics Society.

Crandall, B., Klein, G. A., and Hoffman, R. R. (2006). *Working Minds: A Practitioner's Guide to Cognitive Task Analysis*. Cambridge, MA: The MIT Press.

Eggleston, R. G. (2003). Work-centered design: A cognitive engineering approach to system design. *Proceedings of the Human Factors and Ergonomics Society 47th Annual Meeting* (pp. 263–267). Santa Monica, CA: Human Factors and Ergonomics Society.

Elm, W. C., Potter, S. S., Gualtieri, J. W., Easter, J. R., and Roth, E. M. (2003). Applied cognitive work analysis: A pragmatic methodology for designing revolutionary cognitive affordances. In E. Hollnagel (Ed.), *Handbook of Cognitive Task Design* (pp. 357–382). Mahwah, NJ: Lawrence Erlbaum Associates.

Endsley, M., Bolte, B., and Jones, D. (2003). *Designing for Situation Awareness*. Boca Raton, FL: CRC Press.

Kohn, L., Corrigan, J., and Donaldson, M., eds. (1999). *To Err Is Human: Building a Safer Health System*. Washington, DC: Committee on Quality of Health Care in America, Institute of Medicine. National Academies Press.

Militello, L. G., and Hutton, R. J. B. (1998). Applied cognitive task analysis (ACTA): A practitioners toolkit for understanding task demands. *Ergonomics, 41*, 11, 1618–1641.

Vicente, K. J. (1999). *Cognitive Work Analysis*. Mahwah, NJ: Erlbaum.

Woods, D. D., and Roth, E. M. (1988). Cognitive engineering: Human problem solving with tools. *Human Factors, 30*, 4, 415–430.

2

The Role of Cognitive Engineering in Improving Clinical Decision Support

Anne Miller and Laura Militello

CONTENTS

All designers in some sense believe they are taking a "human-centered" approach, but ... their own intuitions of "what the user needs" somehow get in the way.

Roth 1997, p. 250

Introduction

Clinical decision support (CDS) is an increasing part of health information technology (HIT). This specialized software, designed to affect clinician decision making about individual patients at the time these decisions are made (Berner and La Lande 2007; Osheroff et al. 2012), has exploded in recent years. However, despite CDS development as early as the 1950s (Boden 2006), initial introduction into health care (e.g., Mycin) in the 1970s (Buchanan and Shortliffe 1984), some promising findings of improved clinician performance (Garg et al. 2005), and more recent incentives, such as Meaningful Use Stage 2 certification (IoM 2001; DHHS 2012), the adoption of CDS in clinical practice has been tenuous (Kilsdonk et al. 2011; Sahota et al. 2011; Bright et al. 2012; Oluoch et al. 2012; Patterson et al. 2012).

Some of the reported reasons for poor CDS adoption include the intrusive nature of alerts and reminders and lack of integration into clinical work (Sidebottom et al. 2012; Streiff et al. 2012; Tawfik et al. 2012; Wan et al. 2012; Wu et al. 2012). However, details about where to include clinical reminders and why or how workflow integration is lacking are rarely defined or clarified. It is also unclear whether the reported problems reflect issues specific to an individual clinic setting or whether they arise from more general causes. In this chapter, we examine clinical decision making in outpatient environments and explore an approach for better integration of CDS technology into clinical work.

Clinical Decision Making in Practice

In order to ground our examination, we begin by illustrating a typical outpatient clinic visit. In general, patients who attend outpatient clinics are physiologically more stable and their health and disease physiology more predictable than acute or critically ill patients admitted into hospital units.

Our hypothetical physician, Dr. Finlay, is a practicing internal medicine physician who has worked in a large metropolitan hospital for 16 years. Internal medicine involves the diagnosis and management of patients from health to often multidisease processes, for example, chronic type 2 diabetes with cardiovascular complications. Each week, Dr. Finlay sees approximately 30 patients in her clinic. Annually, she attends approximately 1400 patient clinic visits per year for approximately 850 patients from her local area. She has attended more than 20,000 visits over the course of her career.

One of Dr. Finlay's patients is Jennifer Whittaker, an amiable 58-year-old moderately obese woman with type 2 diabetes mellitus that was diagnosed 15 years ago. Jennifer has made an appointment to see Dr. Finlay as her blood sugar levels have become harder to manage. Jennifer also complains of sore knees that feel hot to the touch and that prevent her from walking her dog. Daily dog walks have resulted in Jennifer losing 10 lb. of weight over the last 12 months. In the 15-min consultation, Dr. Finlay first wants to determine Jennifer's current health status, focusing on what might be causing Jennifer's current problems.

Dr. Finlay reviews Jennifer's past and family history: does Jennifer recall close relatives who have had sore knees? Jennifer's mother had severe arthritis, and Jennifer is afraid that she may be getting this debilitating disease. Dr. Finlay also reviews her last clinic visit notes about Jennifer and observes that her plan of care hasn't changed much over the past two years. The sore knees are a new problem. Dr. Finlay confirms that Jennifer's knees look red and swollen and feel quite warm to the touch. Dr. Finlay decides that the diabetes is not Jennifer's main problem today, as her diabetes has only been problematic since Jennifer's knees began to hurt. Thus, if she can address this issue, Dr. Finlay reasons that Jennifer's diabetes will stabilize. Dr. Finlay orders some blood tests, a knee x-ray, and a referral to a rheumatology specialist for further advice and prescribes an anti-inflammatory medicine to provide short-term pain relief.

The Modeling Problem

Dr. Finlay has seen more than 20,000 patients who are like Jennifer. Thus, Dr. Finlay views Jennifer's health problems through a rich experiential base. Based on her examination of Jennifer's problems and her previous exposure to similar patients, Dr. Finlay generates hypotheses and tests them using various exploratory strategies (e.g., blood tests, procedures, the effects of medications on the symptoms) to either confirm or refute her diagnoses. Bennett and Flach (2011) identify this "learning by doing" as an abductive reasoning strategy where knowledge from past experience is applied to specific problems to develop hypotheses about the situation that are then tested through action. Rosen (1991) describes this process from an epistemological perspective as the "modeling relation" (Figure 2.1).

In our example, Jennifer (the patient) is the natural system on the left of Figure 2.1. Arrow "①" represents self-organizing dynamics occurring in the natural system, which are the cause for Dr. Finlay's encounter with Jennifer. Dr. Finlay is the observer system. She builds a mental model of Jennifer's condition by first encoding (arrow "②") information and then by using inferential processes (arrow "③") informed by past experience and formal learning to draw conclusions about Jennifer's problems. Dr. Finlay's inferences are then "decoded" (arrow "④") or tested by actions such as ordering blood tests and x-rays, making referrals, and prescribing medicines. Encoding and decoding functions aim to bring the natural system and its observer's mental

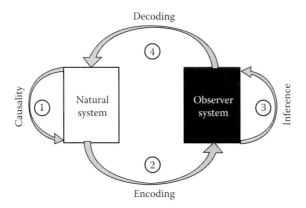

FIGURE 2.1
Rosen's modeling relation (modified).

models into a level of alignment or coincidence. The greater the interaction between the natural world and the observer's mental models, the closer the alignment likely to be.

The modeling relation echoes other models and explanations of human interaction with the world. Brunswik (1943), Simon (1969), and Gibson (1979), and later, Suchman (2007) and Hutchins (1995), each contextualized human cognition/perception/judgment and action within a relative environment and provided a foundation for highly productive systems modeling in sociotechnical environments (Rasmussen 1986; Vicente 1999; Burns and Hajdukiewicz 2004; Mosier 2013). More recently, Bennett and Flach (2011) have developed a model that is similar to Rosen's that includes a technological interface that mediates the natural and observer systems.

In Figure 2.2, Jennifer (the patient) is a representative of the ecology or the problem domain on the left. The CDS/electronic health record (EHR) is an interface medium that lies either fully or partially between the patient ecology and the clinician. Dr. Finlay's mental model is represented on the right as belief, values, and knowledge. Coincidence is more complicated in Figure 2.2 than in Figure 2.1. In addition to the natural and observer systems, coincidence is also required between the interface and the patient (natural system), and the interface and the observer.

Brown (1986) argues that technological systems such as EHRs should be understood as separate systems that co-occur with observer and natural systems. Using Rosen's modeling relation from this perspective, Figure 2.3 shows some of the complexity of the relations among (1) the technologist (i.e., the software designer, business analyst, programmer) and the technology system (i.e., the EHR) including its databases and technical architecture, and

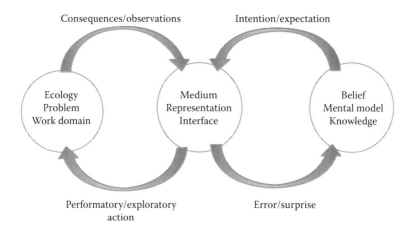

FIGURE 2.2
Bennett and Flach's ecological model. (From Bennett, K.B. and Flach, J.M., *Display and Interface Design: Subtle Science, Exact Art.* Boca Raton, Florida: CRC Press, p. 32, 2011.)

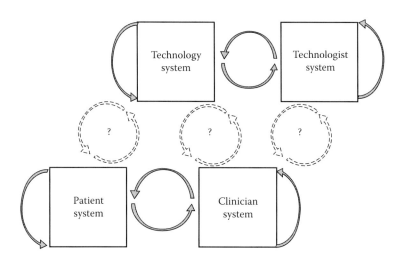

FIGURE 2.3
Extending Rosin's modeling relation to include EHRs.

the laws and relationships that govern the interactions within this interactive system; (2) the clinician and the patient; and (3) the complex relations among these systems.

The goal of the technologist and the clinician in Figure 2.3 is to achieve high levels of coincidence between their mental models and their respective ecological systems (technology; patient). In the interface medium, the technologist wants an accurate representation of the content, rules, and constraints encoded in the technology system; the physician wants an accurate representation of the patient. The overall system objective is to achieve coincidence between the clinician–patient and the technologist–technology subsystems. We propose that many of the issues observed in relation to CDS result from lack of coincidence between the technology–technologist and the patient–clinician systems, as indicated by the dashed arrows in Figure 2.3. We also suggest that cognitive engineering models and methods offer important tools for developing strategies to bridge the gap between the technology and natural systems. The first step is to characterize the problem.

Clinical and Computerized Decision Making

Bridging the gap between the technologist–technology and the patient–clinician systems begins with a more detailed analysis of the relation between these systems. HIT systems have two remarkable characteristics: they have a vast capacity to store information, and they are capable of executing complex computational formalisms and logic rules rapidly with little or no error. Without these characteristics, probability science and large-scale complex multivariate analyses, for example, could not have advanced. However, from a decision perspective, the fundamental change in adopting HIT systems has been to reduce the time, data storage limits, and computational constraints that previously bounded rational human decision making (Simon 1991; Gigerenzer et al. 1999; Todd et al. 2012). With unbounded rationality, new views of the world can be presented in ways that were largely unheard of in previous generations.

However, technologist–technology systems do not interact with patients directly. Technology systems simply store the information and rules fed into them and as such vicariously reflect specific patients and patient populations. There are limitations associated with representing and manipulating vast data stores. One of these is loss of nuance, as patients and clinical work must be represented in computable ways, for example, using standardized language (e.g., SNOMED; see Ruch et al. 2008).

Figure 2.4 is a technologist's view of clinicians' work. Clinical work is conceived as a data retrieval and, in later steps (not shown), as a data entry process. In this view, clinical work is represented as a well-structured, sequential, and thus computable process. Data and variable sets from many thousands of patients are or can be aggregated and transformed into well-defined variable inputs and predicted outcomes. Thus, clinical problems may be solved using algorithmic, mathematical, and rule-based strategies (e.g., evidence-based protocols). In an age when the complexity of clinical work is increasing, many expect the value of these types of tools to also increase. Progress is currently limited in part by our ability to better integrate these computational approaches into clinical practice.

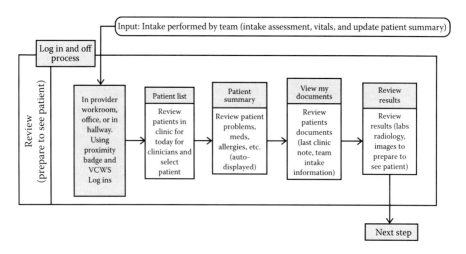

FIGURE 2.4
Software engineer's view of clinical work.

In contrast to the idealized view of clinical work in Figure 2.4, Dr. Finlay's own problem-solving ability is bound by time, available information, and of course, computational capacity (Simon 1969; Gigerenzer et al. 1999; Todd et al. 2012). Dr. Finlay needs to size up the situation quickly and accurately based on the information she has before her, most saliently Jennifer. Jennifer's problems are not well structured. Variables such as knee pain are ambiguous, goals are ill-defined and often competing (Is the primary problem knee pain or diabetes?), and outcomes are loosely and provisionally defined. Variables can express nonlinear relationships and can change dynamically over time. However, Simon (1969) observed that humans are remarkably good at making decisions in uncertain environments using nonalgorithmic strategies that involve optimal, though not optimized, solutions. He also observed that humans rarely use formal (computational) logic to solve real-world problems, preferring instead to use heuristic rules of thumb.

Although research has highlighted human vulnerabilities in decision making under uncertainty (Kahneman et al. 1982), recent studies (Todd et al. 2012) using ecologically valid examples suggest that people can be at least as good as and in some cases better (faster, more accurate) problem solvers than computer-based solutions. When situations and environments are familiar, bounded strategies such as satisficing (Simon 1969) and fast and frugal heuristics (Gigerenzer et al. 1999), including recognition (Zsambok and Klein 1997; Klein 1998), are highly effective. Despite the dominance of these forms of decision making in our daily lives, analytical and computational approaches are favored in CDS. Some heuristics (e.g., the frequency of medications prescribed for certain categories of patients or in certain clinics) are gaining currency in CDS applications; however, guidelines about how to represent computed heuristics are limited.

Cognitive engineering is an approach that can be used to design systems and technologies that support naturalistic decision making as opposed to prescriptive or optimized decision processes. We suggest that cognitive engineering is underrepresented in health care CDS literature. The challenge for cognitive engineering, in collaboration with health information technologists, is to develop methods and approaches to better align computational and clinical decision support.

Effects of CDS on Clinical Decision Making

In this section, we examine the effects of some CDS implementations on clinical decision making as a means for identifying areas for further research and development in cognitive engineering. In particular, we explore prompts and reminders as one of the most commonly used decision support modalities.

Prompts and Reminders

Prompts and reminders are rule- or criteria-based messages designed to alert clinicians to some part of a process that may otherwise be omitted or erroneously executed (Figure 2.5). They may be presented passively, a visual indicator for example, or actively using more intrusive pop-up dialog boxes. They may be instructive, providing information designed to educate or inform (e.g., that a vaccination may be due), or may be directive, designed to inform clinicians about a critical step or potential quality/safety breach (e.g., a drug–allergy incompatibility).

The Veterans Health Administration was one of the earliest prompts and reminders implementers. For each patient, a tailored list of reminders appears on a computerized patient record system cover sheet. Each reminder links to a dialog box that prompts clinicians to enter specific information and then offers a set of actions. For example, clinicians can easily document symptoms and actions using radio button selections and free text fields. Tests, consults, and medications can be ordered directly from the dialog box. Entered data and actions are automatically pasted into the progress note. Although seemingly elegant, when implemented in clinical settings, a number of barriers arise (Saleem et al. 2005).

FIGURE 2.5
Veterans Affairs Medical Center computerized clinical reminder for colorectal cancer screening.

- *The CDS becomes a burdensome task rather than a support for core clinical activities.* In some clinical settings, so many reminders were developed that the list became unwieldy and intrusive. Some reminders target routine activities such as the use of sunscreen and wearing seatbelts; others aid in managing chronic conditions such as reminding physicians that the patients are due for a diabetic foot exam. Screening reminders target depression, substance abuse, and posttraumatic stress disorder and have been used to increase the administration of flu and pneumonia vaccines. Reminders were also developed to aid research data collection. Reminders were not grouped or prioritized in the reminders list.

  Context-independent lists like these are often seen as ancillary tasks to be shed when things get busy rather than as tools to ensure that important elements of care would not be omitted (Saleem et al. 2005). Reminding patients to use sunscreen and alerting clinicians about potentially adverse drug interactions are undoubtedly important, but their importance is context dependent. It may, for example, be more appropriate to direct sunscreen reminders to patients: (a) in patient websites as a public health advisory during summer months and (b) to patients who have sunscreen-relevant problems such as skin cancer. Alternatively, it may be just as effective to use posters or videos in the waiting area or exam room to increase awareness of the importance of sunscreen, reserving reminders for addressing chronic conditions and preventative care issues requiring clinical intervention. Selecting which reminders to display where and how should not depend on what is possible but rather should be prioritized by clinical value and informed by the context of the visit.

- *CDS does not always accommodate variations in clinical workflow* (Ash et al. 2004; Cook et al. 2004; Militello and Klein 2013). For some providers, limiting or altogether eliminating time interacting with the computer during the patient exam is a priority. Some providers choose to review the patient record before the exam and to document after the patient has left in order to better establish rapport and to focus on direct interaction with the patient. The reminders, however, were designed to be used during the exam and include prompts to obtain information from the patient. If the provider did not remember to obtain the exact information required by the reminder, he/she would be left with the option of either leaving the reminder unresolved, in which case it would continue to stay on the list until the missing data were added, or "estimating" to fulfill the reminder.

  This is an example of a tendency to design software that requires the user to follow an optimal or idealized process. In fact, the actual process is often determined in real time by a range of variables including access to equipment (including computers), which staff are available, how much time is available, what the patient needs, etc. Effective decision support must include enough flexibility that the user can realize the intended benefits even as the workflow changes to accommodate both routine and unexpected changes in the work context.

These two barriers to the effective use of clinical reminders are illustrative of the types of problems documented in the HIT literature. Table 2.1 presents a list of other documented barriers to effective CDS use. It is interesting to note that many of these barriers are consistent with those identified by Reisman (1996) in a review of CDS technologies nearly 20 years ago.

- *Decision strategy mismatches.* Clinical decision making is a complex process that requires different skills and types. As in Jennifer's case, understanding why a patient is visiting a clinic, what the patient's primary and current problems are (Jennifer's blood sugar levels or

TABLE 2.1

Barriers to Effective CDS Use

	References
Requires a level of focused attention not suitable for a highly interruptive context resulting in • Increase in data entry errors • Increase in orders entered on behalf of the wrong patient	Ash et al. (2004)
Increases workload resulting in • Time burden • Cognitive burden • Loss of cognitive focus on clinical work • Fragmentation • Shift focus from efficacy to completeness	Ash et al. (2004); Saleem et al. (2005)
Misrepresents work as linear resulting in • Decreased ability to deviate from routine sequences • Barriers to addressing urgent problems without completing less urgent steps • Work-arounds: users must devise strategies to circumvent CDS in order to provide appropriate care	Patterson et al. (2002); Ash et al. (2004); Saleem et al. (2005)
Degraded communication/coordination including • Computer entry replaces conversation • Loss of feedback • Reduced ability to detect errors	Patterson et al. (2002); Ash et al. (2004); Saleem et al. (2005)
Decision support overload such as • Too many reminders and alerts	Ash et al. (2004); Saleem et al. (2005)
Information presented in a manner that violates expectancies • Humans must adapt a decision process to determine what the technology is "thinking"	Patterson et al. (2002); Saleem et al. (2005)
Changes in priorities such as • Technologically monitored activities may take the highest priority during goal conflicts	Patterson et al. (2002)

her sore knees), and the context around them (Jennifer's ability to walk her dog) requires judgment and knowledge acquired through experience with Jennifer and with patients like her. For a more in-depth case study highlighting decision-making challenges in health care, see the work of Wears (2009).

Deciding a course of action can often depend on elements beyond the therapeutic option's demonstrated efficacy including the patient's access to available therapies, ability to pay for therapies, support system, and priorities and values. Thus, decisions about a course of care involve the patient and the clinician weighing and negotiating options and planning their implementation. They depend on the ability to aggregate, synthesize, or build a picture of the patient's current situation and to project a course of care.

These types of highly contextualized and goal-oriented decisions involve experiential and knowledge-based processes (Rasmussen 1986), which include recognition and pattern matching, signal detection, and prioritization (Klein 1998) in a specific context. Without careful integration into clinical workflow, rule-based decision strategies such as those represented in prompts and reminders may not be helpful in supporting these types of decisions and instead may be intrusive and interruptive.

After the patient's situation has been assessed and a course of action has been decided, computerized tools are needed to check the validity of the action plan as it is executed and implemented. Safe and effective care delivery depends on checks for patient allergies, potentially adverse interactions, or out-of-range medication doses critical to patient safety. Prompts and reminders that draw attention to dependencies and alert to inconsistencies or incompatibilities such as potentially adverse drug-to-drug interactions, drug–allergy interactions, or out-of-range or inappropriate dosing are critical last defenses that protect against adverse events.

- *CDS can hinder the development of expertise.* In addition to disrupting situation assessment and awareness processes during a clinic visit, researchers have argued that rule-based applications applied to context-dependent tasks requiring judgment and decision making can interrupt the acquisition of important cognitive skills (Bainbridge 1981, 1983, 1989, 1992; Sheridan 1997; Klein 2003). Of course, in some cases, physicians simply choose not to use poorly designed CDS, but as CDS becomes more tightly integrated into EHRs and quality measures are linked to compliance with CDS, even poorly designed CDS can have an impact. Some of the features of CDS technologies may actually hinder the development of expertise. A common CDS strategy is to filter large amounts of data to present only key pieces of information and important evidence-based recommendations for action. Although presenting actions without the associated data to support them may reduce cognitive load, this approach may severely limit the user's ability to examine the data for trends, outliers, and important contextual information that may aid future recognition. With no practice at examining the entire data sets, users may never develop the ability to build and maintain an assessment of the situation; rather, they will be forced to rely on computer-based recommendations.

Similarly, decision support technologies may inadvertently place decision makers in a passive mode. Rather than actively seeking data from a number of sources (i.e., the patient, other members of the health care team) and drawing on his/her own experience to determine typicality or an anomaly and actively interpret the situation, the user will rely on the decision support to assess, interpret, and recommend action. If the decision support technology does not let the user see its rationale or reasoning, the pedigree of the data may be invisible. The user is unable to determine which data are estimated versus known, which set of guidelines or rules are driving the recommended actions, and which data are driving a particular recommendation. This opacity may discourage users from developing their own mental models. If the rule-based decision support makes it difficult to deviate from a prescribed process, it may discourage users from adapting based on situational factors, again hindering them in building an experience base in which they actively build an assessment and determine an appropriate action based on that assessment (Klein 2003).

Cognitive Engineering and HIT

Cognitive engineering has been applied across a range of complex industry and technology systems including nuclear power plants, aviation, and military systems. See the work of Hoffman and Militello (2008) for a discussion of the evolution of cognitive engineering approaches. In spite of the increasing use of cognitive engineering to create effective decision support and training in complex domains (Cooke and Durso 2010), applications in health care are relatively recent and few (but see Chapters 3 and 4 of this book for specific analysis and design-oriented examples). In this chapter, we highlight four ways in which cognitive engineering can help to bridge the gap between the technology system and the clinical system.

First, cognitive engineering provides insight into the goals at both a system level (i.e., provide high-quality cost-effective patient-centered care) and at an individual level (i.e., determine how to prioritize and manage Jennifer's symptoms). Articulating goals, particularly in challenging situations, provides an important foundation for the design. Furthermore, an understanding of user goals allows for more meaningful usability testing as design concepts are instantiated in prototypes. Too often, usability tests are conducted in a decontextualized way. Participants are asked to work through a series of tasks without realistic problems and goals to inform actions. Findings from this type of usability testing may uncover basic issues with the user interface design, but without scenarios that include realistic problems and goals, it is difficult to assess the potential impact of specific design concepts on the cognitive work necessary to achieve goals at all levels of the system.

In one recent study, for example, researchers used a decision-centered design (a specific approach to cognitive engineering; Militello and Klein 2013) to inform usability testing for a series of modules on the Indian Health Service's EHR, the Resource and Patient Management System (RPMS). Experienced users of RPMS were asked to complete a series of goal-driven tasks using prototype designs. The findings included an assessment of how well each module aided the users in developing and maintaining accurate mental models, obtaining the information they needed to make timely and appropriate decisions, reducing common errors, and rapidly scanning for relevant information. These relatively rapid usability tests revealed important considerations to be addressed before a final version of each module was released (Lopez et al. 2012). Other studies have been conducted to improve EHRs (Berg 1999; Ash et al. 2004), bar coding in medication administration (Patterson et al. 2002), and computerized physician order entry (Pirnejad et al. 2008; Peute et al. 2010).

Second, cognitive engineering provides insight into the information needs of operators. An overemphasis on algorithm-based decision support limits information presented to data that are computable. Designers of algorithm-based decision support seek innovative ways to reduce information in the environment to computable elements. Cognitive engineering, in contrast, seeks to identify the information needs required to complete key tasks in challenging situations—regardless of the form of that information. By examining the information needs, information currently available, and information gaps, designers are able to provide support for more complex macrocognitive skills.

This is particularly important in health care where clinicians must respond to novel events, make important assessments and judgments with incomplete and ambiguous information, and manage conflicting goals and priorities. Many of the complex tasks health care professionals must accomplish can be informed by base rates and probabilities, but that is only one piece of the puzzle. Cognitive engineering methods are designed to highlight how complex macrocognitive skills are used within a specific domain. These methods highlight how skilled operators make difficult decisions, make sense of challenging situations, plan and dynamically replan based on changing conditions, coordinate and maintain common ground within a team and across an organization, detect problems early enough to intervene, manage risk and uncertainty, and manage attention (Klein et al. 2003). For example, a series of studies of emergency department status boards (Nemeth et al. 2004; Wears and Perry 2007; Wears et al. 2007) and trauma center whiteboards (Xiao et al. 2007) highlighted the critical role these seemingly straightforward information-sharing artifacts play in collaboration, task management, team attention management, task status tracking, task articulation, multidisciplinary problem solving and negotiation, resource planning and tracking, socialization, and team building. Xiao (2005) has more broadly examined the role of artifacts and collaborative work in health care.

Third, cognitive engineering brings a set of methods designed for just such complex environments. Interview and observation methods, knowledge representation methods, and qualitative data analysis methods have been refined and tailored based on research and design activities in nuclear power (Rasmussen 1986), military command and control (Klein et al. 1997), aviation (Endsley 2003), military logistics (Eggleston 2003), and many other complex domains.

Perhaps the most well-known and widely applied cognitive engineering method is cognitive task analysis. Cognitive task analysis methods are used to understand complex work via semistructured interviews and ethnographic observations. Cognitive task analysis interview methods tend to be incident-based in that interviewees are asked to relate a challenging lived experience. These interview techniques are designed to aid experienced practitioners in articulating tacit knowledge, including goals, strategies, and critical cues that contribute to skilled decision making, judgments, assessments, and sense making. Decisions are represented in the context of workflow using a range of knowledge representation techniques including vignettes and decision requirement tables. This increases the likelihood that the resulting designs will address real problems and that they will be usable in a range of real-world contexts. See the work of Crandall et al. (2006) for a thorough discussion of cognitive task analysis methods.

Although applications of cognitive task analysis in health care have been somewhat rare historically, they are becoming more common. Some examples include the use of cognitive task analysis methods to explore cognitive challenges associated with complex tasks including neonatal intensive care unit nursing (Crandall and Getchell-Reiter 1993; Militello and Lim 1995), laparoscopic cholecystectomy (Dominguez et al. 2004), information management in computerized physician order entry (Weir et al. 2007), and colorectal cancer tracking and management (Lopez and Militello 2012).

Fourth, cognitive engineering brings a perspective or a sensibility to the role of humans in a complex system. Cognitive engineering approaches advocate for an understanding of the work context and the challenging decisions humans must make. This is in contrast to algorithmic approaches that tend to define a process that an operator should (or in some cases, must) comply with in order to use the software successfully. Cognitive engineering approaches, in contrast, seek to develop decision support that provides key information in a way that is compatible with human decision-making strategies—allowing the operator to follow or deviate from standard processes as required by the situation.

The recognition-primed decision model (Klein 1998), for example, describes an intuitive decision strategy that experts commonly employ to make decisions in the face of complexity. Rather than using analytic strategies such as generating and comparing multiple options, experienced personnel in time-pressured high-stakes environments are much more likely to recognize a situation as familiar and to implement an acceptable course of action without lengthy deliberation and analysis. In most cases, these recognition-primed decisions are made rapidly, smoothly, and very effectively. Cognitive engineering approaches focus on supporting these adaptive human capabilities, because the human part of any system is generally responsible for stepping in and compensating for technological limitations (Woods et al. 2010). In complex systems, humans are tasked with noticing anomalies, detecting problems, and determining what actions to take to avoid negative outcomes. Poorly designed decision support technologies can actually hinder these important functions by requiring adherence to a predefined process.

In spite of the impressive contributions cognitive engineering has made to our understanding of work and the role of HIT in supporting or hindering decision making and other cognitive skills, a few examples exist in which cognitive engineering principles have been applied to real-world widely used CDS technologies. This is due, in part, to the separation of the cognitive engineering community and the health informaticists

In recent years, a promising trend toward cross-pollination from these communities has emerged. An increasing number of health informaticists have become interested in cognitive engineering and attended relevant workshops. This brief exposure, however, is generally not enough to fully integrate and to effectively apply cognitive engineering methods and principles. Conversely, more HIT development teams are seeking cognitive engineering support. All too often, however, this type of consulting is sporadic and not integrated well enough into the design and development effort to realize the full benefits of cognitive engineering. For cognitive engineering to truly make a difference, we—and others (Scott et al. 2005)—advocate for truly multidisciplinary project teams that include health informaticists, software developers, and cognitive engineers throughout the entire front-end analysis, design, and development process. A multidisciplinary project team supports the development of a common ground across all team members. This approach allows the informaticists and the software developers a better view of the end user and the complexity of the work, and allows the cognitive engineer a better view of the constraints those who design and develop the software must accommodate.

Conclusions

The next important step for HIT is to better integrate algorithmic and naturalistic approaches to decision support. We contend that cognitive engineering can help to bridge the gap between the technology system and the patient–clinician system (Figure 2.4). Cognitive engineering offers a framework for understanding goals at multiple levels of the system; a focus on information needs required to complete tasks in challenging situations; cognitive task analysis methods for understanding a first-person perspective with regard to macrocognitive skills such as decision making, sense making, dynamic replanning, and others; and a perspective that emphasizes supporting human cognition over an enforcing and "optimal" predefined process. This emphasis on work in real-world contexts provides insight into benefits as well as the drawbacks of algorithmic approaches to decision support. A more thorough and accurate understanding of clinical work allows designers to prioritize decision support features based on real needs rather than based on what data are available and computable. Furthermore, cognitive engineering methods can be used to identify mismatches between the technology system and the patient–clinician system, highlighting important leverage points for improvement.

References

Ash, J.S., Berg, M., and Coiera, E. (2004). Some unintended consequences of information technology in health care: The nature of patient care information system-related errors. *Journal of the American Medical Informatics Association*, 11(2), 104–112.

Bainbridge, L. (1981). Mathematical equations or processing routines? In Rasmussen, J. and Rouse, W.B. (Eds.), *Human Detection and Diagnosis of System Failures*. NATO Conference Series III: Human Factors, Vol. 15, pp. 259–286. New York: Plenum Press.

Bainbridge, L. (1983). Ironies of automation. *Automatica*, 19, 775–779.

Bainbridge, L. (1989). Development of skill, reduction of workload. In L. Bainbridge and S.A. Ruiz-Quintanilla (Eds.), *Developing Skills with Information Technology*, pp. 8–116. New York: Wiley.

Bainbridge, L. (1992). Mental models in cognitive skill: The case of industrial process operation. In Y. Rogers, A. Rutherford, and P. Bibby (Eds), *Models in the Mind*, pp. 119–143. London: Academic Press.

Bennett, K.B., and Flach, J.M. (2011). *Display and Interface Design: Subtle Science, Exact Art*. Boca Raton, FL: CRC Press.

Berg, M. (1999). Patient care information systems and health care work: A sociotechnical approach. *International Journal of Medical Informatics*, 55(2), 87–101.

Berner, E.S., and La Lande, T.J. (2007). Overview of clinical decision support systems. In Berner, E.S. (Ed.), *Clinical Decision Support Systems: Theory and Practice* (2nd Ed.). New York: Springer.

Boden, M.A. (2006). *Mind as Machine: A History of Cognitive Science* (Vols. 1–2). Oxford, UK: Clarendon Press.

Bright, T.J., Wong, A., Dhurjati, R. et al. (2012). Effect of clinical decision support systems: A systematic review. *Annals of Internal Medicine*, 157, 29–43.

Brown, J.S. (1986). From cognitive to social ergonomics and beyond. In Norman, D.A., and Draper, S.W. (Eds.), *User Centered Design*. Mahwah, NJ: LEA.

Brunswik, E. (1943). Organismic achievement and environmental probability. *Psychological Review*, 50, 255–272.

Buchanan, B.G., and Shortliffe, E.H. (1984). *Rule-Based Expert Systems: The MYCIN Experiments of the Stanford Heuristic Programming Project*. Reading, MA: Addison-Wesley.

Burns, C.M., and Hajdukiewicz, J.R. (2004). *Ecological Interface Design*. Boca Raton, FL: CRC Press.

Cook, R.I., Nemeth, C., and Brandwijk, M. (2004). Technical work studies: Understanding human work amid complexity, uncertainty, and conflict. In *Administration for Healthcare Research and Quality 3rd Annual Patient Safety Conference*. Arlington, VA.

Cooke, N.J., and Durso, F. (2010). *Stories of Modern Technology Failures and Cognitive Engineering Successes*. Boca Raton, FL: CRC Press.

Crandall, B., and Getchell-Reiter, K. (1993). Critical decision method: A technique for eliciting concrete assessment indicators from the intuition of NICU nurses. *Advances in Nursing Science*, 16, 42–51.

Crandall, B., Klein, G., and Hoffman, R.R. (2006). *Working Minds: A Practitioner's Guide to Cognitive Task Analysis*. Cambridge, MA: MIT Press.

Department of Health and Human Services (DHHS) (2012). Medicare and Medicaid Programs: Electronic Health Record Incentive program Stage 2 (Final Rule). RIN0938-AQ84. Available at http://www.himss.org/content/files/2012-21050_PI.pdf.

Dominguez, C.O., Flach, J.M., McDermott, P.L., McKellar, D.M., and Dunn, M. (2004). The conversion decision in laparoscopic surgery: Knowing your limits and limiting your risks. In Smith, K., Shanteau, J., and Johnson, P.E. (Eds.), *Psychological Investigation of Competence in Decision Making*. Cambridge: Cambridge University Press.

Eggleston, R.G. (2003). Work-centered design: A cognitive engineering approach to system design. In *Proceedings of the Human Factors and Ergonomics Society 47th Annual Meeting* (pp. 263–267). Santa Monica, CA: Human Factors and Ergonomics Society.

Endsley, M.R. (2003). *Designing for Situation Awareness: An Approach to User-Centered Design*. New York: Taylor & Francis.

Garg, A.X., Adhikari, N.K.J., McDonald, H. et al. (2005). Effects of computerized clinical decision support systems on practitioner performance and patient outcomes: A systematic review. *Journal of the American Medical Association*, 293(10), 1223–1238.

Gibson, J.J. (1979). *An Ecological Approach to Visual Perception*. New York: Taylor & Francis Group.

Gigerenzer, G., Todd, P.M., and The ABC Research Group (1999). *Simple Heuristics That Make Us Smart.* New York: Oxford University Press.

Hoffman, R., and Militello, L.G. (2008). *Perspectives on Cognitive Task Analysis: Historical Origins and Modern Communities of Practice.* New York: Taylor & Francis.

Hutchins, E. (1995). *Cognition in the Wild.* Cambridge, MA: The MIT Press.

Institute of Medicine (IoM) (2001). *Crossing the Quality Chasm: A New Health System for the 21st Century.* Washington, DC: National Academy Press.

Kahneman, D., Slovic, P., and Tversky, A. (1982). *Judgment under Uncertainty: Heuristics and Biases.* Cambridge, UK: Cambridge University Press.

Kilsdonk, E., Peute, L.W., Knijnenburg, S.L., and Jaspers, M.W. (2011). Factors known to influence acceptance of clinical decision support systems. *Studies in Health Technology and Information*, 169, 150–154.

Klein, G.A. (1998). *Sources of Power: How People Make Decisions.* Cambridge, MA: MIT Press.

Klein, G. (2003). *Intuition at Work.* New York: Doubleday.

Klein, G., Kaempf, G.L., Wolf, S., Thorsden, M., and Miller, T. (1997). Applying decision requirements to user-centered design. *International Journal of Human-Computer Studies*, 46(1), 1–15.

Klein, G., Ross, K.G., Moon, B.M., Klein, D.E., Hoffman, R.R., and Hollnagel, E. (2003). Macrocognition. *IEEE Intelligent Systems*, 18(3), 81–85.

Lopez, C.E., and Militello, L.G. (2012). Colorectal cancer screening decision support: Final report (unpublished technical report, Contract No. 200-2011-M-41884).

Militello, L.G., and Klein, G. (2013). Decision-centered design. In Lee, J.D., and Kirlik, A. (Eds.), *Oxford Handbook of Cognitive Engineering* (pp. 261–271). Oxford: Oxford University Press.

Militello, L.G., and Lim, L.S. (1995). Patient assessment skills: Assessing early cues of necrotizing enterocolitis. *Journal of Perinatal and Neonatal Nursing*, 9(2), 42–52.

Mosier, K.L. (2013). Judgment and prediction. In Lee, J.D., and Kirlik, A. (Eds.), *The Oxford Handbook of Cognitive Engineering*. New York: Oxford University Press.

Nemeth, C., O'Connor, M., Cook, R., Wears, R., and Perry, S. (2004). Crafting information technology solutions, not experiments, for the emergency department. *Academic Emergency Medicine*, 11(11), 1114–1117.

Oluoch, T., Santas, X., Kware, D. et al. (2012). The effect of electronic medical record-based clinical decision support on HIV care in resource constrained settings: A systematic review. *International Journal of Medical Informatics*, 81, 83–92.

Osheroff, J.A., Teich, J.M., Levick, D. et al. (2012). *Improving Outcomes with Clinical Decision Support: An Implementer's Guide* (2nd Ed.). Chicago: HIMSS.

Patterson, E.S., Cook, R.I., and Render, M.L. (2002). Improving patient safety by identifying side effects from introducing bar coding in medication administration. *Journal of the American Medical Informatics Association*, 9(5), 540–553.

Patterson, S.M., Hughes, C., Kerse, N., Cardwell, C.R., and Bradley, M.C. (2012). Interventions to improve the appropriate use of polypharmacy for older people. *Cochrane Database of Systematic Reviews*, 5, CD008165.

Peute, L.W., Aarts, J., Bakker, P.J., and Jaspers, M.W. (2010). Anatomy of a failure: A sociotechnical evaluation of a laboratory physician order entry system implementation. *International Journal of Medical Informatics*, 79(4), e58–e70.

Pirnejad, H., Niazkhani, Z., van der Sijs, H., Berg, M., and Bal, R. (2008). Impact of a computerized physician order entry system on nurse–physician collaboration in the medication process. *International Journal of Medical Informatics*, 77(11), 735–744.

Rasmussen, J. (1986). *Information Processing and Human-Machine Interaction: An Approach to Cognitive Engineering*. New York: Elsevier Science Publishing Co.

Reisman, Y. (1996). Computer-based clinical decision aids. A review of methods and assessment of systems. *Informatics for Health and Social Care*, 21(3), 179–197.

Rosen, R. (1991). *What Is Life?* New York: Columbia University Press.

Ruch, P., Gobeil, J., Lovis, C., and Geissbuhler, A. (2008). Automatic medical encoding with SNOMED categories. *BMC Medical Informatics and Decision Making*, 8(Suppl 1), S6.

Sahota, N., Lloyd, R., Ramakrishna, A. et al. (2011). Computerized clinical decision support systems for acute care management: A decision-maker-researcher partnership systematic review of effects on process of care and patient outcomes. *Implementation Science*, 6, 91.

Saleem, J.J., Patterson, E.S., Militello, L., Render, M.L., Orshansky, G., and Asch, S.M. (2005). Exploring barriers and facilitators to the use of computerized clinical reminders. *Journal of the American Medical Informatics Association*, 12(4), 438–447.

Scott, R., Roth, E.M., Deutsch, S.E. et al. (2005). Work-centered support systems: A human-centered approach to intelligent system design. *IEEE Intelligent Systems*, 20(2), 73–81.

Sheridan, T.B. (1997). Task analysis, task allocation, and supervisory control. In Helander, M.G., Landauer, T.K., and Prabhu, P. (Eds.), *Handbook of Human-Computer Interaction*, 2nd ed., pp. 87–105. Amsterdam: Elsevier Science.

Sidebottom, A.C., Collins, B., Winden, T.J., Knutson, A., and Britt, H.R. (2012). Reactions of nurses to the use of electronic health record alert features in an inpatient setting. *Computers, Informatics Nursing*, 30, 218–226.

Simon, H.A. (1969). *The Sciences of the Artificial* (3rd Ed.). Cambridge, MA: MIT Press.

Simon, H.A. (1991). Cognitive architectures in a rational analysis: Comment. In VanLehn, K. (Ed.), *Architectures for Intelligence*, pp. 25–39. Mahwah, NJ: Erlbaum.

Streitt, M.B., Carolan, H.T., Hobson, D.B. et al. (2012). Lessons from the Johns Hopkins multi-disciplinary venous thromboembolism prevention collaborative. *British Medical Journal*, 344, e3935.

Suchman, L.A. (2007). *Human-Machine Reconfigurations: Plans and Situated Actions* (2nd Ed.). New York: Cambridge University Press.

Tawfik, H., Anya, O., and Nagar, A.K. (2012). Understanding clinical work practices for cross-boundary decision support in e-health. *IEEE Transactions on Information Technology in Biomedicine*, 16, 530–541.

Todd, P.M., Gigerenzer, G., and The ABC Research Group (2012). *Ecological Rationality: Intelligence in the World*. New York: Oxford University Press.

Vicente, K.J. (1999). *Cognitive Work Analysis: Towards Safe, Productive and Healthy Computer-Based Work*. Mahwah, NJ: Lawrence Erlbaum Associates.

Wan, Q., Makeham, M., Zwar, N.A. et al. (2012). Qualitative evaluation of a diabetes electronic decision support tool: Views of users. *BMC Medical Informatics and Decision Making*, 12, 61.

Wears, R.L. (2009). What makes diagnosis hard? *Advances in Health Science Education*, 14(1), 19–25.

Wears, R.L., and Perry, S.J. (2007). Status boards in accident and emergency departments: Support for shared cognition. *Theoretical Issues in Ergonomics Science*, 8(5), 371–380.

Wears, R.L., Perry, S.J., Wilson, S., Galliers, J., and Fone, J. (2007). Emergency department status boards: User-evolved artefacts for inter- and intra-group coordination. *Cognition, Technology and Work*, 9(3), 163–170.

Weir, C.R., Nebeker, J.J.R., Hicken, B.L., Camp, R., Drews, F., and LeBar, B. (2007). A cognitive task analysis of information management strategies in a computerized provider order entry environment. *Journal of American Medical Informatics Association*, 14(1), 65–75.

Woods, D.D., Dekker, S., Cook, R., Johannesen, L., and Sarter, N. (2010). *Behind Human Error*. Surrey, UK: Ashgate Publishing.

Wu, H.W., Davis, P.K., and Bell, D.S. (2012). Advancing clinical decision support using lessons from outside of healthcare: An interdisciplinary systematic review. *BMC Medical Informatics and Decision Making*, 12, 90.

Xiao, Y. (2005). Artifacts and collaborative work in healthcare: Methodological, theoretical, and technological implications of the tangible. *Journal of Biomedical Informatics*, 38(1), 26–33.

Xiao, Y., Schenkel, S., Faraj, S., Mackenzie, C.F., and Moss, J. (2007). What whiteboards in a trauma center operating suite can teach us about emergency department communication. *Annals of Emergency Medicine*, 50(4), 387–395.

Zsambok, C. E., and Klein, G. (Eds.). (1997). *Naturalistic Decision Making*. New York: Psychology Press.

3

Team Cognitive Work Analysis as an Approach for Understanding Teamwork in Health Care

Catherine M. Burns

CONTENTS

Introduction

Health care environments are complex with diverse teams of professionals from diverse backgrounds working on hard problems. In understanding the cognitive work of health care, one must look at the challenges of the domain as well as the distribution of work across the teams involved. With this view in mind, we evolved traditional cognitive work analysis (CWA) and extended it to provide explicit guidance on team interactions. This chapter provides an introduction to team cognitive work analysis (TCWA) and explains how it was used in the context of understanding interactions in a labor and delivery department.

TCWA was developed to provide a team emphasis and lens on CWA, and should be viewed in this way and not as an independent or different analysis. CWA (Vicente 1999) is a five-level analysis that examines work domain structure, decision-making depth, strategies, social–organizational factors, and worker competencies. CWA in its traditional form does not exclude the analysis of team situations. However, the TCWA discussed here modifies CWA to focus attention on team-relevant requirements. In TCWA, each level closely follows Vicente's terminology and concepts for CWA (Vicente 1999). However, the advantage of TCWA (Ashoori and Burns 2010, 2013) is that it examines teamwork with deliberate intent, seeking to identify explicitly teamwork constraints in the four different phases of CWA: (1) team work domain analysis (team WDA), (2) team control task analysis (team ConTA), (3) team strategy analysis (team StA), and (4) team competencies analysis (team WCA). In TCWA, these analyses are deepened to add an additional perspective of explicitly identifying team constraints. Since the analysis of teamwork is a component of the social, organizational, and cooperation analysis of CWA, one way of looking at TCWA is as a contributor to this analysis phase.

Team CWA

While TCWA as we have drawn it together is relatively recent, the application of CWA to teams is not new. Many authors have used CWA in the past to explore team interactions (e.g., Burns and Vicente 1995; Burns et al. 2005, 2009). TCWA builds on much of this past work. In particular, the contextual activity template of Naikar et al. (2003, 2006) and Naikar (2006) and the distributed CWA views provided by Jenkins et al. (2008a,b) are key techniques that can be used in a TCWA. TCWA takes team views of four phases of CWA: work domain analysis (WDA), control task analysis (CnTA), strategies analysis (StA), and worker competencies analysis (WCA). The general approach to these phases is discussed in the following.

WDA (Team View)

A WDA is a hierarchical functional expression of the work domain and its structure. By hierarchical, a WDA starts with purposes and objectives at the highest level and works to components and object and their attributes at the lowest level of the tree. By functional, a WDA seeks to explain why and how the domain works without explicitly describing operator actions or tasks. This is analogous to the idea of a functional specification, which specifies performance but not how a design should achieve that performance. By taking this approach, WDA explores the space where actors must operate and identify complexities, components, and designed-in assumptions of the environment without specifying the potential actions they might take. The specification of actions and strategies is not ignored in CWA but occurs in other phases of CWA. In a team view of WDA, one can look at how purposes are shared or conflict across the sections of the domain, what components are shared (or not), and what components must move between actors. Team members may have shared goals, or individual goals, and may need to coordinate the use of equipment or resources. Understanding the distribution of the domain and the domain control across the team provides a first indication of what team members may need to have appropriate situation awareness of the joint efforts of the team.

CnTA (Team View)

A CnTA identifies information processing activities involved in key system tasks. The tasks that are often of most interest are start-up or planning tasks, diagnosis or problem-solving tasks, and shutdown or discharge tasks. These are typical "control tasks" since they represent a set, or a class of tasks, with essentially the same control goals within the class. In addition, a CnTA can be used to explore the differences in behavior, identifying the shortcuts and the heuristics that experts may use to operate more efficiently. In a Team CnTA, the emphasis shifts to task distribution as we look at activities across multiple actors, which activities occur synchronously or asynchronously, and information artifacts that must be passed between actors. The control tasks remain the same as these relate to classes of tasks in operating the system. But by looking at the distribution of control tasks, or information flows across the team, adding the team view of CnTA provides a layer of richness on top of the CnTA.

StA (Team View)

An StA examines the different ways that actors can achieve their goals with a system. An StA may look at the differences between novice and expert actors, high- and low-workload actors, or actors with different kinds of training. In a Team StA, the structure of the team can be included as well as looking at when and how the team may reconfigure to meet certain task demands. It may also look at the different strategies of new and more experienced teams.

WCA (Team View)

A WCA looks at the skills, rules, and knowledge (SRK; see Rasmussen 1983) that each actor must bring to fulfill his or her role and operate effectively. While the SRK provides a rich description of expertise and required levels of cognitive control, this is only one potential view of worker competence. A key addition to this from the team view comes from looking at the social competencies that an actor must develop. As an example, actors in different roles may require different social skills. In particular, the social competencies of team leaders can have a strong influence on team performance. Just as certain rules and knowledge enable different levels of cognitive control, different social competencies enable communication and coordination within a team. Where the other TCWA models build layers on CWA models, the social competencies model provides a uniquely different component than the SRK.

In the section "Example of TCWA," we will show how TCWA can be applied in a health care setting.

Example of TCWA

A labor and delivery unit at a large city hospital can be a busy place. While many patients enter for routine deliveries, the situation can change to an emergency situation very quickly. One obstetrician may be managing several patients or a few depending on the day. If an emergency begins to occur, a surgical team must be assembled quickly, bringing in a new set of expertise. Before and after delivery, different personnel may be involved. For example, a pediatrician or a postsurgical team may become involved. Two key participants move through the work domain, interacting with all teams and units: the mother and the baby. As is commonly seen in health care work domains, the mother participates not only as the key object in the problem-solving space but also as a decision-maker and stakeholder.

Observations

In total, 31 h of observations were conducted in the labor and delivery unit at a 1000-bed tertiary care hospital (Ashoori and Burns 2011; Ashoori et al. 2014). The research protocol was reviewed and received ethics approval from both the University of Waterloo and the Research Ethics Board of the hospital where the observations occurred. Observations included admitting patients, performing surgery if needed, and transferring to the recovery room. The interactions of the Caesarean-section (C-section) surgical team were analyzed because of the tight team coordination required of this team. These teams consist of several smaller teams such as the anesthesia team, the nursing team, the pediatric team, and the obstetrical team. The teams are used to working with each other and are formed for the 45–90 min surgical events. The team members' expertise ranged from novice to expert.

From the observations, TCWA models were developed. These are discussed in order from team WDA, team ConTA, team strategies, and team social competencies.

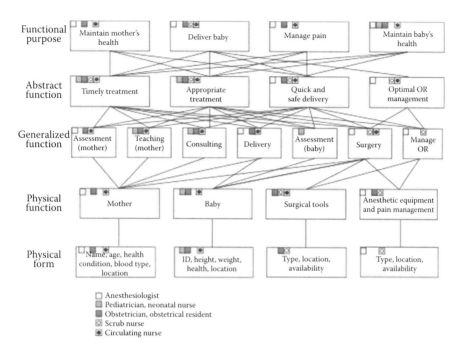

FIGURE 3.1
WDA showing team member regions of control.

Team WDA

A WDA reveals constraints at different levels of abstraction in the work domain. The key difference in a team WDA is that we look at who is influenced by which constraints and what constraints are shared. The team WDA in Figure 3.1 shows the domain constraints for the labor and delivery unit, which are distributed across five key stakeholders. The WDA components and connections remained the same in the team WDA; however, the team WDA distributes the components across different stakeholders. From this analysis, we can see that while many values and purposes are shared, there are potential conflicts when functional objects must be shared and processes must be coordinated. A key example of this is the fact that the mother is accessed by several different people, and the baby is checked by a set of sometimes different people. Similarly, the mother's health record must be accessible to different people and for different purposes. There are conflicting values as well. For example, a quick delivery may not necessarily be the safest form of delivery or set the team up for the most effective pain management approach.

Team ConTA

Team ConTA can reveal information flow between team members during key tasks, making it one of the most practical analyses in a TCWA. We examined different control tasks in various routine situations and built the decision ladders (Rasmussen et al. 1994) to analyze what needs to be done for each task. For example, Figure 3.2a shows the decision ladders for newborn evaluation in a routine situation. The pediatric team and the circulating nurses contribute to this control task. The steps are numbered for simplicity. The shaded boxes show the decision ladder elements that are activated in the situation. The baby's arrival (step 1 in the figure) is a signal for the pediatric team to immediately start the assessment process (step 2). The outcome of this activity is a set of measured data (step 3). Based on the collected information, the pediatric team identifies whether the baby is healthy or needs special care (step 4). In a routine situation, when the baby is healthy, the pediatric team knows that they need to document the results of the baby assessment (step 5); they formulate the procedure to complete the assessment (step 6) and identify the sequence of actions to perform (step 7). After that, the pediatric team and the circulating nurses are ready to implement the actions (step 8). The ladders in Figure 3.2 are typical of a regular ConTA. In an abnormal situation where a procedural solution is not a good fit, the decision makers may have to reason between different options before

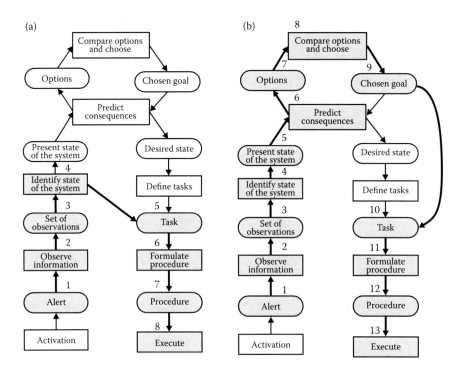

FIGURE 3.2
Control task analysis in two situations: routine tasks (a) and abnormal situations (b).

choosing a course of action. Figure 3.2b shows the decision ladder in case of an emergency. By using this approach, one can clearly see that the emergency situation takes the team from routine procedures and requires the team to diagnose a more complex situation and evaluate available options, all while under time stress. While this is useful, the decision ladder alone does not show the roles and actions of the various team members involved in the tasks.

Team ConTA improves on the decision ladder by showing interactions between team members through the decision wheels (Figure 3.3). Each wheel shows a team with each team member comprising a portion of the wheel. The decision ladder of each team member is drawn within the slices, and the connections between the ladders represent the interactions between the team members. Synchronous and asynchronous activities are highlighted as well as communication flows between the teams and the team members. The links are numbered for simplicity. Similar to Figure 3.2, the typical decision ladder, once the baby has arrived, the pediatric team starts the initial observation to make sure that the baby is healthy (link 1). The circulating nurses help to complete the baby assessment (link 3). The circulating nurses share the observation task with each other (link 4), and then they plan the sequence of actions to complete that task (link 5). After completing the observation, the circ-1 nurse updates the pediatrician with the requested information (link 6), and the pediatric team decides if the baby needs some special care (link 2). The decision wheel allows individual decision ladders to be displayed, showing team member roles. The wheels show the decision steps taken by teams as a unit. Interactions between individuals and teams can be shown. The overall depiction of team work is much richer than in a decision ladder alone. In contrast to other approaches with multiple decision ladders, the decision wheel is more scalable.

While the decision wheels are a good representation to show how different parties interact on a single control task, the contextual activity template (CAT; Naikar et al. 2003) is a good representation to show how individuals are involved in multiple control tasks or work together on tasks in various situations. For these reasons, we consider the CAT to be another useful representation within TCWA. Figure 3.4 shows a version of a CAT for the surgical team. Work situations are shown along the horizontal axis, and roles and responsibilities are shown along the vertical axis. The circles are placed to indicate the involvement of the role in the workplace situation. In particular, the CAT clearly shows the coordinated involvement of several people in work situations such as surgeries and patient preparation.

Team ConTA can show the various task steps in different control tasks, where complexities lie, and, through the decision wheels, how teams coordinate their activities and timing to accomplish joint tasks.

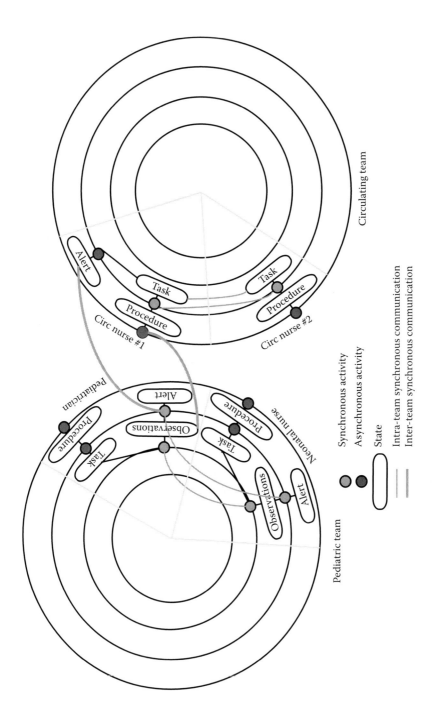

FIGURE 3.3
Decision wheel showing control task elements distributed across actors and synchronicity.

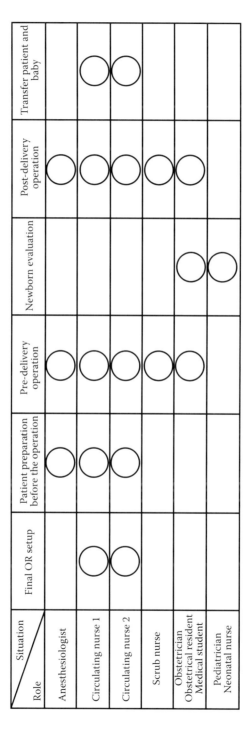

FIGURE 3.4
CAT for the team.

Team Strategy Analysis

Strategies can be analyzed in several different ways. In this context, operating effectively as a team was important. For the team strategy analysis, we built an information flow map to examine routine and emergency situations in a pediatric case as there are distinctly different work patterns in these two situations. This is shown in Figure 3.5.

We also looked at coordination strategies in the routine and emergency situations. In Figure 3.6, we have shown the coordinative structure in both routine situations and in an abnormal emergency situation. In the figure, the routine situation is shown on top, and the abnormal emergency situation is shown below. In the emergency situation, there is tighter coordination and stronger central coordination through the emergency pediatric team leader. In this configuration, the looser team structure has been replaced by a tighter coordinative structure enabled by the addition of the emergency pediatric

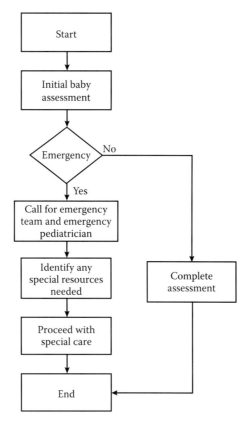

FIGURE 3.5
Information flow map showing routine and emergency situations.

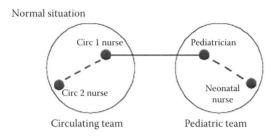

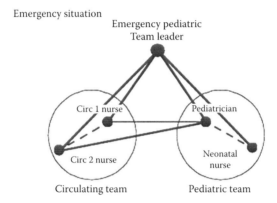

FIGURE 3.6
Structural changes that occur when the team faces an abnormal emergency situation.

team leader to the team. He or she communicates, directs, and interacts with all four members of the team. The pediatrician has also taken a stronger role, interacting directly with the second circulating nurse rather than communicating through the first nurse and expecting him or her to manage interactions with the second nurse.

Competency Analysis

In a regular CWA, the SRK framework is used to assess the competencies that each worker should possess (Rasmussen 1983; Vicente 1999). A TCWA supplements this by examining the social competencies that are equally as important in having an effective team. Social competencies are distinctly different from SRK. In particular, team leaders must be able to communicate, motivate, and instill confidence. Other team members must be able to listen, mitigate conflict, and maintain team communication. While there are many sources for team skills and personality types, Belbin (1981) is a commonly used reference. Table 3.1 presents the SRK inventory, and Table 3.2 presents the social competencies for potential roles of the labor and delivery unit teams.

TABLE 3.1

SRK Inventory

Team Member	Skill-Based Behavior	Rule-Based Behavior	Knowledge-Based Behavior
Pediatric team: —Pediatrician —Neonatal nurse	An experienced pediatric team should be able to quickly identify the emergency situations and if the baby requires special care.	The pediatric team should be able to look up information available in the OR such as care algorithms.	Once the required information is collected, the pediatric team should be able to analyze supplementary information and make a decision about the criticality of the situation.
Circulating team: —Circ-1 nurse —Circ-2 nurse	The experienced nurses should be aware of the set of measurements required for baby assessment in the OR.	The circulating team should be able to look up information such as care algorithms to determine equipment and other resources required.	The circulating team should be able to identify a list of signs and symptoms to observe.

TABLE 3.2

Social Competency Inventory

Team Member	Functional Role	Team Role	Social Skills Required	Technical Skills Required
Pediatric team	Pediatrician	Coordinator specialist	Single-minded, self-starting, dedicated to provide knowledge and skills in rare supply Confident; a good chairperson; should be able to clarify goals, promote decision making, and delegate well	Specialized; four or five years of residency after graduation from medical school is required
	Neonatal nurse	Team worker	Cooperative, perceptive, and good interpersonal skills; should be able to listen, build, and avert friction while being assertive in their role	Expert nurse with specialty training in neonatal nursing

(continued)

TABLE 3.2 (Continued)

Social Competency Inventory

Team Member	Functional Role	Team Role	Social Skills Required	Technical Skills Required
Circulating nurse	Circulating nurse 1	Coordinator	Confident; a good chairperson; should be able to clarify goals, promote decision making, and delegate well	The trained nurse who has passed the training period required for participating in a C-section surgery
	Circulating nurse 2	Team worker	Cooperative, perceptive, and good interpersonal skills; should be able to listen, build, and avert friction while being assertive in their role	The trained nurse who has passed the training period required for participating in a C-section surgery

Discussion

To understand team work in health care, there are many dimensions of cognitive work that need to be considered. We have proposed TCWA as one possible structure approach for capturing team requirements. TCWA builds on a CWA approach but looks specifically at team interactions. In particular, the main reasons for taking the TCWA approach are to identify the following:

1. *Shared and individual situation awareness needs.* The team WDA view begins by mapping out the workspace, all components, and competing purposes. By identifying which components and purposes are shared in the work domain, the situation awareness needs for one or more people can be identified. Shared purposes show when those people should be coordinating together, and shared components show what objects in the domain will be touched by more than one actor. Purposes that are not shared by two actors require special attention. If these conflict, there is a potential for team conflict and ineffectiveness as team members work against each other to achieve competing goals. Display support that shows the objectives of the other team member and their individual contributions to a group goal may help to improve this situation.

2. *Security and change tracking.* When work domain objects are touched by multiple users, there are several consequences. First, one user may make changes of which other users are unaware. Unknown

changes may cause unexpected actions in the future. The Team WDA can identify objects that are shared between more than one user. These objects may need additional design attention to indicate changes and historical state, alert other users, or secure them from further changes.

3. *Improving communication flow.* Understanding which stages are synchronous or asynchronous in the CnTA can help improve team communication. Synchronous stages need to be supported by collaborative tools, whereas asynchronous stages may require time tracking, notification, and historical tracking tools.

4. *Building better teams.* When social competencies are considered as well as skill rules and knowledge, more effective team members can be developed. Adding social competencies acknowledges that team members can have the right skills and knowledge, but with poorly matched social or leadership skills, they will fail to be effective. Understanding that teams reconfigure into different structures can also help to understand how team members may play different roles.

Conclusions

Health care environments are rich and dynamic with rapidly changing requirements. To be able to design for these environments in a way that is resilient to the change requirements, and compatible with cognitive work, a systematic cognitive engineering approach is required. TCWA is one approach that has been designed for these environments. In this chapter, we explained CWA and demonstrated how CWA can be used to examine interactions in a labor and delivery unit of a hospital.

Acknowledgments

This chapter acknowledges, summarizes, and reinterprets work from the studies conducted by Maryam Ashoori, with assistance from Kathryn Momtahan and Barb d'Entremont of the Ottawa Hospital, without whose collaboration this work could not have occurred. The author thanks the Natural Sciences and Engineering Research Council of Canada for supporting this work through a Discovery Grant Accelerator Award.

References

Ashoori, M., and Burns, C.M. (2010). Reinventing the wheel: Control task analysis for collaboration. *Proceedings of the Human Factors and Ergonomics Society's 54th Annual Meeting*, 274–278.

Ashoori, M., and Burns, C.M. (2011). Control task analysis in action: Collaboration in the operating room. *Proceedings of the Human Factors and Ergonomics Society's 55th Annual Meeting*, 272–276.

Ashoori, M., and Burns, C.M. (2013). Team cognitive work analysis: Structure and tasks. *Journal of Cognitive Engineering and Decision Making*, 7, 123–140.

Ashoori, M., Burns, C.M., Momtahan, K., and d'Entremont, B. (2014). Using team cognitive work analysis to reveal healthcare team interactions in a labour and delivery unit. *Ergonomics*, 57, 973–986.

Belbin, R.M. (1981). *Management Teams: Why They Succeed or Fail*. Oxford: Butterworth-Heinemann.

Burns, C.M., and Vicente, K.J. (1995). A framework for describing and understanding interdisciplinary interactions in design. *Proceedings of the 1st Conference on Designing Interactive Systems: Processes, Practices, Methods, & Techniques*, 97–103.

Burns, C.M., Bryant, D., and Chalmers, B. (2005). Boundary, purpose, and values in work-domain models: Models of naval command and control. *Systems and Humans, IEEE Transactions on Systems, Man and Cybernetics*, 35, 603–616.

Burns, C.M., Torenvliet, G., Chalmers, B., and Scott, S. (2009). Work domain analysis for establishing collaborative work. *Proceedings of the Human Factors and Ergonomics Society's 53rd Annual Meeting*, 314–318.

Jenkins, D.P., Stanton, N.A., Salmon, P.M., Walker, G.H., and Young, M.S. (2008a). Using cognitive work analysis to explore activity allocation within military domains. *Ergonomics*, 51, 798–815.

Jenkins, D.P., Stanton, N.A., Walker, G.H., Salmon, P.M., and Young, M.S. (2008b). Applying cognitive work analysis to the design of rapidly reconfigurable interfaces in complex networks. *Theoretical Issues in Ergonomics Science*, 9, 273–295.

Naikar, N. (2006). An examination of the key concepts of the five phases of cognitive work analysis with examples from a familiar system. *Proceedings of the Human Factors and Ergonomics Society's 50th Annual Meeting*, 447–451.

Naikar, N., Pearce, B., Drumm, D., and Sanderson, P.M. (2003). Designing teams for first-of-a-kind, complex systems using the initial phases of cognitive work analysis: Case study. *Human Factors*, 45, 202–217.

Naikar, N., Moylan, A., and Pearce, B. (2006). Analyzing activity in complex systems with cognitive work analysis: Concepts, guidelines, and case study for control task analysis. *Theoretical Issues in Ergonomics Science*, 7, 371–394.

Rasmussen, J. (1983). Skills, rules, and knowledge; signals, signs, and symbols, and other distinctions in human performance models. *IEEE Transactions on Systems, Man, and Cybernetics*, 13, 257–266.

Rasmussen, J., Pejtersen, A.M., and Goodstein, L.P. (1994). *Cognitive Systems Engineering*. New York: Wiley.

Vicente, K.J. (1999). *Cognitive Work Analysis, Toward Safe, Productive, and Healthy Computer-Based Work*. Mahwah, NJ: Lawrence Erlbaum Associates.

4

Cognitive Engineering Design of an Emergency Department Information System

Theresa K. Guarrera, Nicolette M. McGeorge,
Lindsey N. Clark, David T. LaVergne, Zachary A. Hettinger,
Rollin J. Fairbanks, and Ann M. Bisantz

CONTENTS

Introduction

The role of information technology (IT) within health care is growing at a rapid rate. When designed effectively, health IT has the potential to improve patient care, safe patient outcomes, and the efficiency with which health care is provided (Institute of Medicine Committee on Quality of Health Care in America 2001; Aspden et al. 2004); however, simply implementing technology does not guarantee these benefits (Littlejohns et al. 2003). Instead, as the number of electronic systems in health care continues to grow, there is a need to ensure that systems are designed based on a nuanced and comprehensive understanding of the complex needs of multiple types of users. For instance, the "end users" of health IT systems include clinical staff (nurses, physicians, pharmacists, technicians, etc.) who enter and view patient health information during direct patient care, administrative workers such as coders and billers, administrators who may be interested in high-level summaries or status overviews, and patients and families who are affected by the information presented.

Emergency departments (EDs) present a unique challenge for information system design. The often quoted 1999 Institute of Medicine (IoM) report (Kohn et al. 1999) indicated that patients in the ED setting had the highest rate of preventable adverse events (Brennan et al. 1991). Subsequent IoM reports have highlighted the inherent risks in EDs and other critical care settings (IoM 2006). Contributing challenges unique to the ED system include the lack of an established physician–patient relationship and limited access to complete medical records. As a result, physicians may not be familiar with the patient's medications, past medical history, or allergies. In addition, patient acuity levels are at their highest, and time-sensitive workflows necessitate a higher prevalence of verbal orders, all while staff are treating multiple patients at once, with frequent interruptions (Chisholm et al. 2000). ED and hospital overcrowding also contribute to the high-risk environment in the ED, as ED patient volumes continue to increase while the total number of EDs in the United States decreases. Together, these factors point to the need for appropriately design IT innovations that can increase efficiency and safety in the emergency medicine setting.

Methods in cognitive engineering (CE) are designed to understand complex information needs and have often been used in other high-risk, complex, safety critical domains. This approach can be used to better understand what information needs are required to care for patients in the setting of the ED. This in turn will allow us to better design and evaluate the interaction of health IT, clinical environments, and the people who are using them. In particular, one CE method called the cognitive work analysis (CWA) can be used to identify information needs in complex sociotechnical systems characterized by complex interacting processes, multiple team members, and critical decisions (Vicente 1999; Bisantz and Roth 2007).

In this chapter, we discuss the results of a CE analysis conducted of an ED (specifically a work domain portion of a CWA) and describe how this analysis identified information needs and informed interface display designs for a novel emergency department information system (EDIS). This research provides a new example of how CE can be applied in the health care domain.

Background

The Emergency Department and Health Information Technology

Prior to computerization, ED patient-tracking systems typically consisted of large manual whiteboards used to track patients' progress throughout their stay in the ED. Providers and staff used the whiteboards to organize and temporarily record medical and demographic information about the patients as well as logistical information about the hospital and ED. However, increased integration of electronic systems has resulted in these manual patient-tracking systems being replaced by electronic EDISs. The new information systems electronically store and present patient information, replacing the manual status boards that were commonly used for managing clinical work. An EDIS may or may not be linked with other health IT (e.g., electronic health record, computerized physician order entry, ambulatory offices, etc.), and users may encounter different levels of integration and interoperability among different software programs.

Current EDISs often mimic the look and layout of the original whiteboards, with patient data arranged in rows and columns; they allow information to be accessed at multiple locations and provide automated record keeping and reporting. However, research indicates that they also impose new constraints on use, may fail to support the work of the health care providers, and introduce new sources of error (Fairbanks et al. 2008). For instance, a study of transitions from manual to electronic status boards identified problems including the following: inflexibility of the electronic system to match data entry methods of caregivers, critical information was "hidden" from view, and physician developed methods for tracking their own workflow were unavailable (Bisantz et al. 2010).

Advancements in technology may come at a cost if not properly designed. Health IT must support the users of the system and the work they perform. Without a full understanding of the type of work and users of a system, the implications of design decisions cannot be fully appreciated. If systems are designed to accommodate administrative functions without considering the implications across all users, then the design of patient care functions may be inadvertently hindered. The effects of new software on staff performance, safety, and patient care must be understood before, during, and after the implementation process, or else flawed designs may have negative patient care consequences or may even be abandoned by users (Berg 2003; Stead and Lin 2009).

Cognitive Engineering

Methods in CE allow the analyses of complex work environments such as health care, and emergency medicine more specifically, from two complementary perspectives: understanding the domain- and situation-based factors (such as competing system purposes, available processes, and limited resources, information, and time) that support, constrain, and otherwise shape performance; and understanding the knowledge, skills, experience, and strategies that experienced practitioners bring to bear on these challenges in order to achieve successful system performance (Bisantz and Roth 2007). Outcomes of CE analyses support the design of information systems that enhance successful work practices (rather than disrupt), and allow system operators to respond appropriately to the diverse and unpredictable events in their environment.

As noted above, understanding the complexities and challenges posed by the work domain within which practitioners operate is a key component of CE analysis. This approach has been used to analytically evaluate a wide range of complex sociotechnical systems, including military and process control systems (Burns et al. 2000; Vernon et al. 2002; Bisantz et al. 2003). Work domain analysis is one component of CWA noted above (Roth and Bisantz 2013). It is used to identify information about the structure of the work domain that both defines the normal functioning of a system and constrains possible activities of humans or automated agents within the domain. Work domain analysis is advantageous for complex system analysis because the work domain, or the system to be controlled, is described separately from users' tasks, goals, events, or automation and therefore is independent of the particular method, strategy, or set of activities agents choose to accomplish their tasks. Instead, this type of analysis can identify functional constraints on actions that users can select, such as what equipment is available or what resources are available to support various functions, by defining performance boundaries and event-independent representations of the system (Rasmussen et al. 1994; Vicente 1999).

Typically, a work domain analysis done within the CWA tradition utilizes an abstraction hierarchy (AH) model that combines both means–ends and part–whole hierarchies to describe system components (Bisantz and Vicente 1994). The means–end hierarchy provides multiple representations of the system, at different levels of physical abstraction, ranging from the high-level purposes of the system to abstract constraints and relationships often resulting from first principles or societal values, to system processes, to functions implemented to achieve those processes, to the physical resources available to accomplish the functions, including their locations and states. The part–whole hierarchy, which is orthogonal to the means–end hierarchy, allows the system to be represented at multiple levels of detail—from the complete system, to subsystems, to individual components. Nodes in the model correspond to system components, at multiple levels of abstraction, and can be used to systematically identify information required to support practitioner activities such as system monitoring and control (Burns and Hajdukiewicz 2004; Roth and Bisantz 2013).

Work domain analysis has been applied in the domain of clinical medicine, such as patient monitoring in operating rooms, cardiac care, diabetes management, and neonatal intensive care monitoring (Hajdukiewicz et al. 1998; Sharp and Helmicki 1998; Burns and Hajdukiewicz 2004; Burns et al. 2008) (also see Chapter 3 of this book on work domain analysis of labor and delivery as well as related works by Ashoori and Burns [2013] and Ashoori et al. [2014]). A recent scoping review of examples of CWA in health care (Jiancaro et al. 2013) identified 19 examples where work domain analysis was applied; however, these were primarily focused on patient level monitoring, disease management, or medical device/system design (e.g., such as the design of the electronic health record to help in clinical decision making regarding patient treatment; Edraki and Milgram 2004; Miller et al. 2009). Effken et al. (2011) used work domain analysis to characterize the broad work environment of nurse managers, and Lopez et al. (2010) conducted a work domain analysis of an inpatient ward but focused only on the purpose of fall prevention rather than the management of the ward itself. Rogers et al. (2006) present a work domain model of an electronic status board used in emergency medicine. Thus, despite the recognition that CWA and work domain models, in particular, have utility in the medical domain, a few previous studies have used Work Domain Analyses to model functions across a caregiving system such as a hospital ED. Additionally, the focus of previous Work Domain Analyses has been primarily patient- or device-specific rather than how a caregiving system is controlled and managed.

Methods

Semistructured interviews and focus groups were conducted with 15 emergency medicine physicians (attending physicians and residents), nurses, technicians, and clerical staff. Participants were recruited from a large tertiary care academic medical center. Three emergency medicine physicians involved with the study served as subject matter experts and were also interviewed.

Prior to data collection, all participants were given an informational letter describing the study and provided verbal consent to participate in the session. Participants were interviewed individually or in focus groups of up to three. The session was conducted in person or over the phone, depending on the availability of the participant. At least three research members were present for each session; one acted as moderator, and the remaining members took notes. At the end of the session, participants were compensated for their time with a $10 gift card.

The semistructured interview questions were focused on identifying purposes, processes, functions, and components of the ED work domain necessary for practitioners to successfully complete their work responsibilities (Bisantz et al. 2003). Example interview questions are shown in Table 4.1.

Analysis

Interview notes were compiled and reviewed by the research team as follows in order to create the AH model: Notes were first reviewed to categorize the participants' comments into the most relevant level of the AH. Example

TABLE 4.1

Example Categories and Questions for the Semistructured Interviews

Priorities and constraints	What are some higher-level guidelines or overriding constraints on accomplishing ED goals? (For example, hospital or departmental policy and procedure, insurance policy, clinical guidelines, patient care protocols, standards of care, "unwritten" policy, or expectations.)
Functions and systems	Can you describe how different kinds of systems and resources inside and outside of the ED are used in accomplishing these goals? (For example, computer applications, such as computerized physician order entry, radiology information system, picture archiving and communication system, EDIS, or patient charts.) What typical processes and primary tasks do you participate in in the ED?
Alternative means for achieving function	What if the systems we just talked about were unavailable or broken? How would you deviate from your typical flow? How would the process of patient care be affected? How could you achieve the goals?
Interaction among functions	How do ED functions interact with others within the hospital? Are there situations where the different functions might interact with each other? Might conflict with or constrain each other? In those cases, how would you decide which would take precedence?
Information sharing and communication requirements for coordination of patient care; role of expertise	What kind of communication/transfer of information do you feel is needed between the different roles (nurses, physicians, technicians, etc.)? How do you ensure that everyone maintains the same "big picture"—that the team maintains a shared understanding of the current situation? What about a particularly challenging incident or time in the ED? What knowledge was needed to deal with that situation? How did the work get done?

categories included descriptions about the system, goals, challenges, constraints, guidelines, and resources. Data were sorted by participant role and were analyzed, and then a repeat analysis was conducted with all the data. Drafts of the model were periodically reviewed with the subject matter experts on the research team to ensure the appropriateness and applicability of the AH model being developed.

Resultant Work Domain Model

Five levels of the AH were developed based on the interview analysis, including more than 100 nodes in the top four levels. The following nodes and relationships were described:

- *Functional purpose*: Four main ED purposes were identified over the course of the interviews: providing high-quality care (to both urgent and more routine cases), acting as the gatekeeper to the hospital (i.e., patient admissions through the ED), maintaining financial viability, and providing medical training and education to the clinician and the staff.

- *Abstract function*: Nine major priorities and constraints were identified: ED culture, ethical treatment of people, flows of knowledge, patient's physiologic constraints, time constraints, patient flow, resource demands, information flow, and regulatory accountability and compliance.

- *General processes*: Five general processes were identified: providing patient care, maintaining situation awareness over the ED, coordination and communication, training, and administrative processes.

- *Physical function*: Seven distinct functional groups were identified: personnel, patients, information systems, communication systems, facilities and equipment, expendable supplies, and regulatory guidelines and hospital regulations.

- *Physical form*: Typical nodes at the physical form level describe the status, characteristics, and form of nodes at the physical function level. For instance, in this model, patient demographic information, health information, alerts, location, and status within the care process would be included as representing patient "form." While such data classes were considered in our analysis, an exhaustive set of nodes was not created and therefore is not shown in Figure 4.1.

Figure 4.1 depicts a high-level view of how all of these components are connected and related to one another. For example, one functional purpose identified is "Gatekeeper for hospital," which demonstrates that the path of patient admission to a hospital can be related to the activities and outcomes of the patient's ED visit. For this particular node, the "Gatekeeper for hospital" functional purpose is related to two abstract functions: "Patient flow" and "Resources vs. demands," because capacity and constraints of these categories can influence the timing and manner in which a patient is admitted to the hospital. "Patient flow" is related to the flow of patients into and out of the phases of emergency care and within the hospital. Although the ED accommodates patient surges, departments such as intensive care units are strictly limited by the number of available beds and staffing ratios. Resources may also effect patient-admission decisions. For instance, a lack of open beds in the hospital may force admitted ED patients to "board" in the ED for an extended period of time until the patient can be discharged, or a bed becomes available. At other times, the specific lack of equipment or resources for a patient may prioritize the patient's need to be admitted to the hospital expeditiously and moved to an inpatient room to free up resources in the ED.

Moving down the AH, the "Patient flow" node at the abstract function level is connected to three of the general function level nodes: "Caring for the patient," "Maintain situational awareness over ED," and "Coordination and communication." "Resources vs. demands" is also connected to the same three general function nodes. This general processes level describes

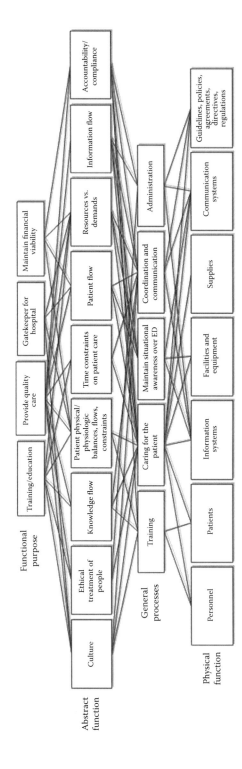

FIGURE 4.1
High-level summary of the ED AH. (Courtesy of Ann Bisantz, University at Buffalo, The State University of New York.)

the processes involved in achieving the goals of the environment (i.e., gateway of the hospital). Patient care may be further separated into subprocesses including triaging, diagnosing, treating, ordering, dispositioning, transporting, and feeding, as shown in Figure 4.2.

"Maintain situational awareness over ED" refers to the need to maintain an understanding of ED functions in terms of available resources, work assignments, and team functioning, as well as the health status of individual patients. Situational awareness allows providers to adequately evaluate equipment utilization and patients' health status. "Coordination and communication" is critical when collaborating patient care between health care team members. Not only does this node represent verbal communication but also includes patient documentation and coordination of staff and schedules. It is important to note that one of the nodes identified at this subprocess level was related to both situational awareness and coordination and communication: team communication and activity monitoring is a characteristic of both nodes at the general function level (see Figure 4.3).

The final two levels of the AH define the physical function and physical form elements. Seven nodes were identified at the physical function level and include personnel; patients; information systems; facilities and equipment; supplies; communication systems; and guidelines, policies, agreements, directives, and regulations. The nodes at this level describe the functional elements of the ED that are needed to achieve the system purposes, and

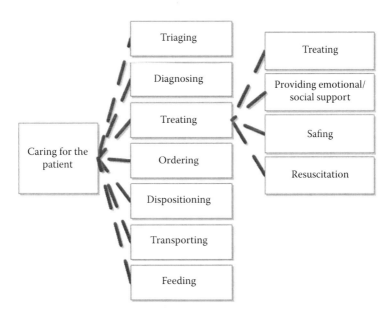

FIGURE 4.2
General process example "Caring for the patient" and subprocesses. (Courtesy of Ann Bisantz, University at Buffalo, The State University of New York.)

FIGURE 4.3
General process examples of "Maintain situational awareness over ED" and "Coordination and communication" and subprocesses. (Courtesy of Ann Bisantz, University at Buffalo, The State University of New York.)

represent the location and the state of the system elements described at the physical function level. Exemplar physical form information was identified such as specific clinical care guidelines; hospital policies, procedures, and guidelines; demographic information from current and previous patient visits; hospital and staff schedules; resources, including descriptions such as the state of the resources; and physical health information about the patient. Connections between nodes across both means–ends and part–whole levels were used during the design phase when information areas and visualization concepts are being brainstormed.

Iterative Display Development

Information based on the AH model was used to develop novel EDIS display concepts, which support the needs of clinicians and staff to work effectively. An iterative design process was implemented consisting of preliminary visualization, conceptualization, and development; internal review and assessment to select a promising subset of visualization concepts; refinement of select visualizations identified from internal review; development of limited function prototypes to support heuristic usability assessment; and iterative revision of the prototype interfaces based on feedback from the usability assessment as well as requirements for implementation in a human-in-the-loop evaluation.

Conceptualization and Development

The first design iteration was primarily concerned with translating the descriptive information in the AH into visualization concepts. Specific information elements of the AH were mapped onto logical display groupings and interface components. For example, a patient's name and bed assignment would need to be displayed together in at least one type of visualization. Traditionally, an EDIS is composed of rows and columns of patient information; specific patient information is filled in across the columns of a particular row for each patient. However, additional information discovered throughout interviews led to the development of separate bed management and status visualizations, as this information also plays a significant role in achieving the providers' goals while working in the ED.

Examples of four bed status visualizations are shown in Figure 4.4a–d, representing typical outputs produced during this phase of the process. In Figure 4.4a, the underlying design concept represents each bed in a list, organized by room number, and indicates whether or not a patient is occupying the bed. Figure 4.4b integrates the bed occupancy visualization with all other types of ED resources. In this instance, the bed is considered to be a resource that is either available or not available for use by a patient. The example shown in Figure 4.4c focuses more on displaying the group of possible beds without reliance on the actual bed numbers. Instead of emphasizing the state of a specific room, this design provides an overall sense of the number of in-use or open beds throughout the ED. Figure 4.4d continues to utilize block shading to indicate bed occupancy, but unlike the previous designs, this visualization arranges the bed information according to the physical layout of the given ED, thus providing additional information that could be used to support staffing (e.g., if staff members are assigned based on areas) or memory for patients (the patient with the heart attack is next to the stockroom).

A total of nine categories of visualizations were created during this initial phase. The groups represented sets of information, as described from the interviews and AH, that would be used in conjunction to accomplish specific goals within the ED:

1. Information about patients waiting for treatment
2. Status of beds
3. Summary view of patients based on the progress of their visit
4. Patient-specific information
5. Staff workload

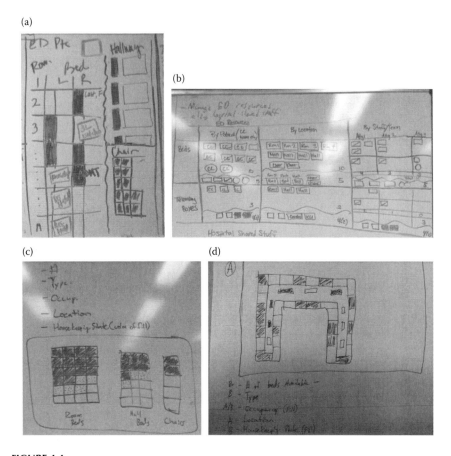

FIGURE 4.4
Four hand-drawn examples for bed status display concepts. Examples include bed status representations that are (a) organized by bed number and type; (b) combined with other ED resources; (c) focused on overall occupancy; and (d) geographic representations. (Courtesy of Ann Bisantz, University at Buffalo, The State University of New York.)

6. Staff tasks

7. Resource and equipment utilization and wait times

8. An overview of the ED function and status

9. Home or main screen from which to view and access all other design areas

Just as multiple designs were created at this stage to represent the status of beds in the ED, a series of designs were created for each of the remaining eight categories listed above. For each category, the information described at the physical form level was identified, and multiple visualization concepts were proposed. Snapshots such as those shown in Figure 4.4a–d were taken of all resulting display concepts.

Initial Internal Reviews and Assessment

In the next phase, more than 35 of the hand-drawn visualizations were selected for further development. Electronic representations of the hand-drawn design concepts were created, and an internal review of the visualization was conducted. Figure 4.5a–d provides a few examples of the visualizations assessed at this stage. Figure 4.5a corresponds with the

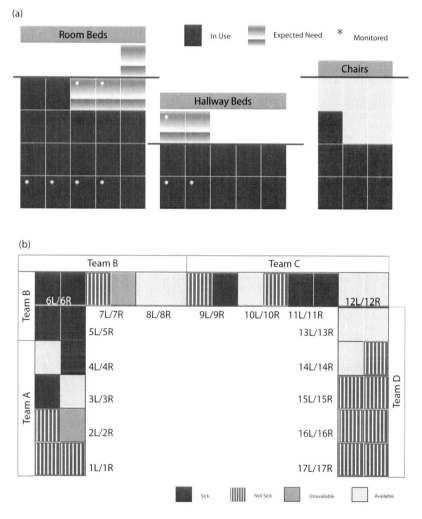

FIGURE 4.5

Four electronic examples of designs evaluated during the internal assessment. Examples include bed status representations: (a) bed status focused on overall occupancy; (b) bed status focused on geographic location of the bed. Text in figures has been altered due to publisher requirements and therefore figures are not identical to those produced by the research team.

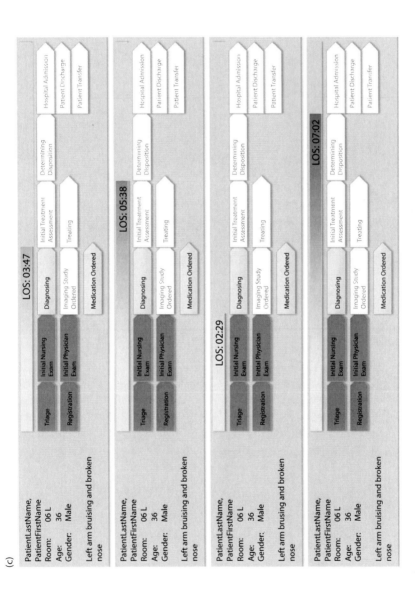

FIGURE 4.5 (Continued)
Four electronic examples of designs evaluated during the internal assessment. Examples include bed status representations: (c) patient-specific information.

(d)

	Wayne, Bruce -26	Kent, Clark -24	Parker, Peter -28
Banner, Robert	10 R 12 R	11 L	
Barton, Clinton	11 R 10 L		13 R 12 L
Frost, Emma		14 R	17 R
Grey, Jean	20 L 18 L		18 R 15 R 15 L 19 R
Gwendolyne, Stacy	07 R	07 L 09 R	08 L 08 R
Kinney, Laura	06 L	04 L 01 L	
Maximoff, Wanda	04 R 03 R	01 R 02 L 03 L 05 R	

FIGURE 4.5 (Continued)
Four electronic examples of designs evaluated during the internal assessment. Examples include bed status representations: (d) staff workload distributions. (Courtesy of Ann Bisantz, University at Buffalo, The State University of New York.)

initial hand-drawn design shown in Figure 4.4c. This design supports a quick understanding of the overall state of all beds. Figure 4.5b corresponds with the sketch shown in Figure 4.4d and shows how a U-shaped ED may appear on the prototype interface. Shading is utilized to indicate the state of the specific beds, and the exact bed location and room number correspond with the position of the square in the diagram. This particular example also overlays the ED staff team assignment for a given set of beds, which may be useful for EDs that assign teams to patients based on treatment location.

Figure 4.5c shows the initial design for a different screen—one that displays patient-specific information and includes basic demographic information, chief complaints, and comments (as is typically found on current ED status board displays). Explicit representations of a patient's length of stay and progress through their ED visit have been added to this display design in order to help clinicians visualize where each patient is in their workup plan and whether the patient is moving through phases in a timely manner.

Figure 4.5d shows a representation for a different area—staff workload. Workload is differentiated based on both the number and complexity of patients assigned to each staff member; the area covered by the bars corresponds to the overall workload and allows comparisons across providers (e.g., to support requests for additional staff to be brought in or to determine who should "pick up" the next upcoming patient). Note that this visualization requires a workload score, corresponding to patient complexity, to be computed—a value that is not currently known. This illustrates an important benefit of the CWA approach: a focus on information needs associated with domain constraints and processes can lead to the specification of novel requirements for information gathering or sensing (Burns et al. 2004)—not just a "repackaging" of what is already known or being collected. In this case, developing a method to assess patient complexity is beyond the scope of the current research but points to a need for further research.

The main objective of this round of review and assessment was to provide all members of the research team (especially those not actively involved in the interface development) with the opportunity to assess the visualizations. Six members of the research team performed the review (four emergency medicine physicians with human factors experience including authors RF and AZ, one emergency medicine nurse, and one human factors scientist) and assessed the interfaces based on the following questions:

- Is anything critical missing?
- Is anything on the display less useful?
- Is anything unlikely to be measurable?
- What tasks might you do with this display?
- When might these tasks be completed with other activities?

Reviewers along with researchers who participated in design (including authors AB, TG, and NM) were organized into two smaller groups to facilitate feedback and discussion. At the end of the small group sessions, the entire team reconvened and reconciled their assessments of the visualizations. Results of this evaluation identified which designs would be used in the subsequent prototype evaluation. One design area, staff tasks, was eliminated from future development after the internal assessment due to limitations of project scope. Additional feedback received from the panel was used to improve the remaining designs before the next stage of assessment. Figure 4.6 shows the resultant design for the bed status screen area. Unlike the previous designs, the new visualization focused on categorizing the beds based on their status and organizing them based on the amount of time in a particular category.

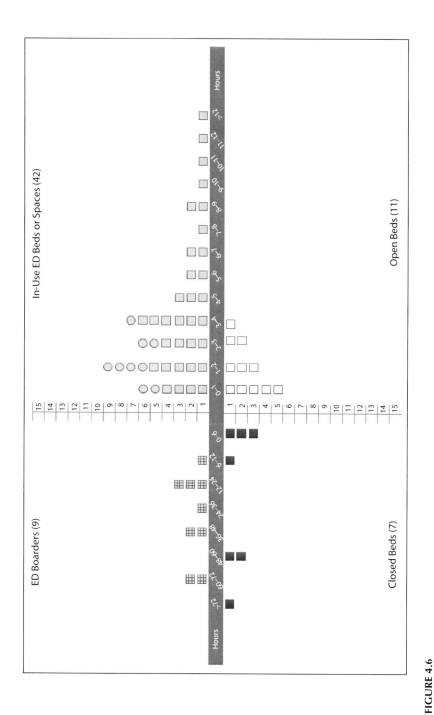

FIGURE 4.6
Revised bed status display used in the usability assessment. (Courtesy of Ann Bisantz, University at Buffalo, The State University of New York.)

Usability Assessment of Prototype Screens

In the next phase of evaluation, emergency medicine physicians and nurses were asked to evaluate the interfaces in a laboratory study. A total of nine physicians and nurses completed the usability assessments. Each participant watched a brief training video, was allowed free time to explore and familiarize himself or herself with the prototype, and then was asked to perform two think-aloud tasks. During the think-aloud tasks, participants provided feedback as to what they were doing to complete the task as well as additional comments about the design of the prototype displays. At the conclusion of the tasks, the participants were also asked to complete two surveys. The first survey asked the participants to evaluate how well the interface helped them complete their tasks, and the second survey asked the participants to assess how usable and how useful the display designs were and to estimate how frequently they would use the displays while working. (The surveys were based on those used by Truxler et al. [2012] to evaluate military operation displays.)

Participants reported generally positive experiences with the interface, and survey results indicated the potential for these visualization concepts and design areas to be helpful in their work environments. For example, a dedicated design area displaying specific information about the patients in the waiting room was a novel concept for all of the participants, and participant ratings and comments were positive for this design area. A full description of the usability assessment and the complete results of this study are reported elsewhere (Clark et al. 2014).

During the session, participants were able to provide qualitative feedback during the think-aloud tasks and any additional comments at the end of each task. During these times, participants provided suggestions based on their individual experiences and preferences. These comments were taken into consideration with all other data collected throughout the research project, and revisions were made in the prototype screens accordingly. Design changes that were recommended at this stage were predominantly recommendations on how to better incorporate the proposed visualizations into their individual workflows and tendencies.

For example, the bed status visualization originally included only ED beds because of the scope of the created AH. The primary focus of the AH was to identify the needs and goals of the ED, and although the specific tasks of assigning patients to ED beds and acting as a gateway to the hospital were both identified during the process, data were not collected at a task-analysis level of detail. However, it was clear after the usability testing that the hospital bed information is a critical piece of information that the ED providers need when admitting patients in order to understand how resource availability in other parts of the hospital affected potential patient flow out of the ED. Figure 4.7 provides an example of the revised display.

Two major design change recommendations made for the overview display were (1) a reallocation of screen space among the different display areas

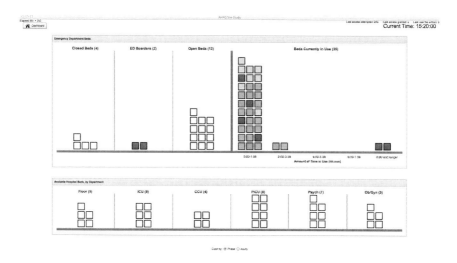

FIGURE 4.7
Final version of the bed status display. (Courtesy of Ann Bisantz, University at Buffalo, The State University of New York.)

and (2) an alternative view of the content presented from the patient-specific information area. Participant feedback supported a greater emphasis on the patient-specific information display than any other individual area when working from the overview display. Participant feedback also indicated that more detailed patient information should be available at the highest interface level and that the length of the patient progress bars without any other information was not informative. For comparison, Figures 4.8 and 4.9 show the overview display before and after these changes.

Discussion

The methods described above provide an example of how information collected through CE methods, specifically, the development of a work domain model, can be combined with methods in user-centered design to produce novel, useful visualizations for health IT systems. Figure 4.10 shows the components of the AH model, which were explicitly represented in the prototype interfaces (shaded nodes), while Table 4.2 provides a summary of the display areas, information represented, examples of the final graphical configuration, and mapping to AH nodes. Note that Appendix A also provides full screen shots of three display areas, which could not be completely represented in Table 4.2. As shown in Figure 4.10, and described in detail in Table 4.2, the prototype displays focus on goals, functions, and processes

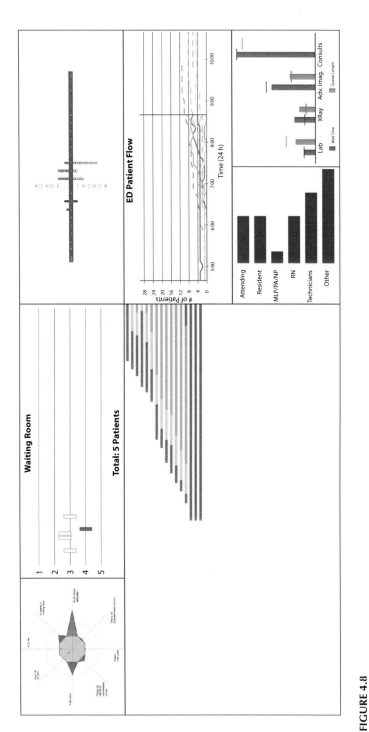

FIGURE 4.8
Overview display used for the first usability assessment. (Courtesy of Ann Bisantz, University at Buffalo, The State University of New York.)

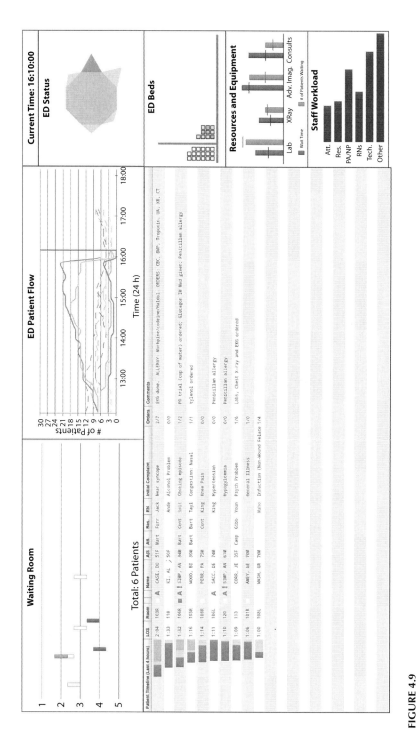

FIGURE 4.9
Revised version of overview display after incorporating usability assessment feedback. (Courtesy of Ann Bisantz, University at Buffalo, The State University of New York.)

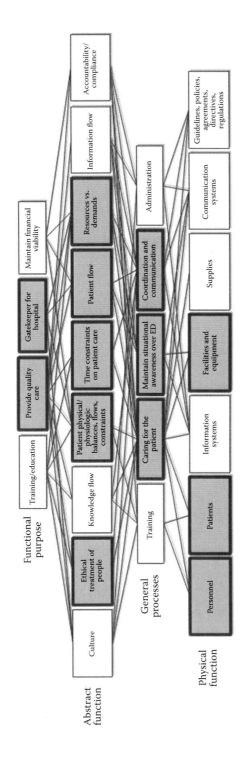

FIGURE 4.10
Highlighted nodes in the AH made up the primary information displayed in the prototype screens. (Courtesy of Ann Bisantz, University at Buffalo, The State University of New York.)

related to providing quality care and serving as the "gatekeeper" or entrance to the hospital. In this way, the displays supported activities related to the management of patient care and patient flow through the ED, and lower-level processes and functions associated with those goals. For example, the displays supported monitoring of the following abstract functions and constraints: ethical treatment (primarily through monitoring timeliness of treatment/appropriate triaging); patient-related constraints; time constraints; flows of patients; and various resource–demand balances (testing, staff, etc.). General processes of caring for the patient, maintaining situation awareness, and communication/coordination were supported through the displays of patient-specific information, phase of care, caregivers, and functionality for communication among caregivers (notes, alerts, and comments) across most of the display areas. Finally, specific information about patient states, personnel workloads, and availability of required laboratory and imagining facilities was also provided in a number of display areas.

One stated benefit of work domain models, such as the AH, is that they can help designers identify important new variables necessary for system monitoring and control: variables that may not have typically been included in previous displays or even variables that are not currently measured (Burns et al. 2004). This analysis was no exception. Variables such as patient pain level and actual vs. typical wait times for imaging studies or laboratory tests may be available but are not typically synthesized, displayed, and used for ED monitoring and assessment. Other variables were developed based on information needs the AH identified. In particular, the concept of specific treatment phase (waiting room, in-bed waiting, assessment and treatment, waiting for reassessment, and dispositioned), which corresponds to indications for task coordination (it is time for someone to recheck the patient) or potential bottlenecks (Why is that patient still in an ED bed when he/she was discharged an hour ago), is novel and was developed through extensive iteration with subject matter experts on the team. Likewise, the concept of acuity ("sick" vs. "not sick" roughly mapping to unstable vs. stable) is a common understanding among ED practitioners but is not typically flagged in ED information systems. Triage scores provide an initial guideline but have a different connotation in terms of urgency depending on the institution and clinical staff. Similarly, the triage scores themselves were annotated with additional information ("a concern") such as patient age or comorbidities in order to provide a finer gradation of presenting severity than the standard triage score provides. Finally, the concept of workload associated with a patient in order to assess overall provider workload is novel within an EDIS and recognizes that patient severity, type of care needed, and phase of care are important components of workload; simply "adding up" assigned patients does not provide a satisfactory method for balancing provider workload. Our system provided a proxy measure of workload based on triage scores. Future research is necessary to identify valid and reliable ways of dynamically measuring patient-associated workload (also see Chapter 9 of this book for a related discussion on nurse workload).

TABLE 4.2

Mapping of Display Area to Nodes at Different Levels of the Abstraction Hierarchy (AH) Model along with the Variables Displayed, Method of Display, and Excerpted Graphical Representation from the Display Area

Display Area Description	AH Mapping	Specific Variables	Example Representation	Display Method
ED Status Provides an overview of how well the ED is performing compared to goals/benchmarks	FP: provide quality care FP: gatekeeper for hospital AF: ethical treatment of people AF: time constraints on patient care AF: resources vs. demands GP: maintain SA over ED	Primary: number of patients in the ED, number of patients in the waiting room; percent of ED patients that are boarders; number of patients that left without being seen; average pain level; time to first medication; time to first doctor; average length of stay. Goal, warning, and unacceptable zones are also provided. Hover or click: Specific variable values and goal ranges		Octagon object display allows quick monitoring of whether variables are in range (all dots within the center green octagon) or out of range (yellow or red points); users can hover to obtain variable values. For each measure, "goal" (green boundary), "alert" (yellow area), and "unacceptable" (outer red octagon) values are determined.
Waiting Room Provides information about patients who are waiting for treatment organized according to severity and waiting time	FP: provide quality care FP: gatekeeper for hospital AF: ethical treatment of people AF: patient physical/ physiologic balances, flows, constraints AF: time constraints on patient care AF: patient flow GP: caring for the patient GP: maintain SA over ED GP: coordination and communication PF: patients	Primary: number of patients waiting, wait time, severity Hover or click: patient name, age, gender, chief complaint, vital signs, triage notes		Patients are represented by colored bars and move left to right as their length of wait time increases. Vertical placement and color of the bar are based on incoming severity. Hovering/clicking on a bar brings up additional details about the patient. Details of up to four patients can be viewed/compared at once. A secondary display area shows information about patients arriving by ambulance/called in by physicians.

Bed Status
Provides information about the number of ED beds that are open, or are not available (due to cleaning, communicability of other patients in the room, etc.). Open beds in other hospital areas are also shown.

FP: provide quality care
FP: gatekeeper for hospital
AF: resources vs. demands
GP: maintain SA over ED
GP: coordination and communication
PF: patients
PF: facilities and equipment

Primary: number of closed, open, or in-use ED beds (including the number of boarders); status of patients in beds (phase or acuity); length of stay of patients in beds; the number of open beds in other hospital units
Hover or click: patient name, age, gender, chief complaint, length of stay, acuity level, phase of treatment, acuity, room number, bed location, bed comment

Beds are represented by open or filled squares, and are grouped according to either hospital floor or ED status category. Beds in the ED, in-use category are further organized by the amount of time the bed has been used by the current patient (e.g., 2-h chunks of time). In-use ED beds are color-coded by the phase of treatment or patient acuity (user selected).

ED Patient Flow
Provides information about patients in each phase of treatment, over time, along with historical patient volumes and projected data

FP: provide quality care
AF: patient flow
GP: maintain SA over ED
GP: coordination and communication

Primary: number of patients in each treatment phase, by time, for previous 24 h; historic data for corresponding time period (e.g., same day, time of day, time of year, etc.); historic data for the next 2 h.
Hover or click: specific values for actual, historic, or projected/historic data

14:00   15:00

Line graphs are used to plot (a) the most recent 24 h of patient volume in each treatment phase and (b) the average or historic data for the same 24 h period of time and the historic data for the next 2 h of time. There is a separate line graph for each treatment phase, and users can toggle between views.

(continued)

TABLE 4.2 (Continued)

Mapping of Display Area to Nodes at Different Levels of the Abstraction Hierarchy (AH) Model along with the Variables Displayed, Method of Display, and Excerpted Graphical Representation from the Display Area

Display Area Description	AH Mapping	Specific Variables	Example Representation	Display Method
Staff Workload Shows the number of patients, patient workload, and total workload assigned to each caregiver	FP: provide quality care AF: resources vs. demands GP: maintain SA over ED GP: communication and coordination PF: personnel	Primary: patient name, workload due to a patient*, assignment of a patient to a caregiver, a patient's acuity or phase of treatment Hover or Click: patient name, acuity, phase of treatment		Patients are represented by horizontally stacked, colored bars. Bar length corresponds to workload associated with a patient. Provider workload is represented by the length of all assigned patient bars and can be compared across providers. Patients are color-coded according to the phase of treatment or patient acuity (user selected). Different types of providers (e.g., physicians, nurses, residents) are shown on different tabs.
Resources and Equipment Present current and historic wait times and the number of people waiting for laboratory tests, imaging orders, and specialty consultations	FP: provide quality care AF: time constraints on patient care AF: patient flow AF: resources vs. demands GP: caring for the patient GP: maintain SA over ED GP: coordination and communication PF: facilities and equipment	Primary: resource type (e.g., x-ray, MRI, Stat Lab, consultant); the number of patients waiting, wait times, historic (typical) number waiting, historic (typical) wait time Hover or Click: specific variable values		Bar graphs are used to present the current number of patients waiting and the amount of time it currently takes the order to be completed for an individual patient. These bar graphs are presented together with a dual y-axis of the number of patients and time. For each bar graph (current data value), the typical value is represented by a line. If the bar graph extends beyond the line, the wait time or the number waiting is more than what is normal for that time, day, etc.

Patient Progress: Overview
Provides summary information about all patients being treated in the ED

FP: provide quality care
AF: patient physical/ physiologic balances, flows, constraints
AF: time constraints on patient care
AF: patient flow
GP: caring for the patient
GP: maintain SA over ED
GP: coordination and communication
PF: personnel
PF: patients

Primary: patient name, age, gender, complaint, length of stay, bed assignment, treatment phase, caregivers, alerts (same name, allergy, medical, nonmedical), vital signs, the number of orders, completed orders, comments
Hover or Click: caregiver notes, alert reasons, full name; click to access detailed patien: progress screen

Length of stay in treatment phases indicated by horizontally stacked colored bars. Total length corresponds to length of stay. Patient details in row/column format, sortable by name, provider, length of stay.

Patient Progress: Detailed
Provides specific details about patient treatment plans, orders, and results

FP: provide quality care
AF: patient physical/ physiologic balances, flows, constraints
AF: time constraints on patient care
GP: caring for the patient
GP: coordination and communication
PF: patients

Primary: patient name, age, complaint, room number, length of stay, vital signs, comments, provider specific notes, alerts, time and type of orders, whether or not orders are completed, whether or not results are abnormal, test results
Hover or Click: test results

Orders (medications, tests, imaging studies, transportation) are shown using a graphical timeline. Unshaded boxes indicate the time and type of order (e.g., "L"—lab test). Shaded boxes indicate a completed order and time of completion. Red boxes indicate an abnormal or concerning result. The color of the timeline background corresponds to the treatment phase. Users can hover over boxes for additional details (e.g., test results). Other display areas on this screen show notes, comments, textual records of order and results, vital signs, patient demographic data, and a link to complete patient chart.

Source: Copyright Ann Bisantz, University at Buffalo, The State University of New York.
* Note that a "workload" score is not easily measureable in the current environment and would need further development to be implemented. The workload score could be determined by a number of factors including initial severity score, number of orders, types of medications, etc.

As noted above, the prototypes we created provided support for activities related to ED management of patient flow and patient care. We did recognize that there are additional purposes related to the training of medical students and residents and maintaining financial viability, and we included these nodes in our AH model for purposes of completeness. However, we did not create prototypes that directly supported these purposes because of the decisions we made regarding the project scope (an important consideration in any CE analysis; see Burns and Bisantz 2008). It would be possible to create additional displays, for instance, which tracked aspects of trainee learning (e.g., the number of procedures completed) or financial accounting, which could be accessed by caregivers and other hospital employees (i.e., education coordinators, financial administrators). Likewise, abstract functions related to knowledge flow (i.e., knowledge gained by trainees) and accountability/ compliance functions were not supported, and management of information flow (that is, support for monitoring of electronic health systems) was considered out of the scope of the project. Constraints due to culture are interesting to consider. It may be challenging to explicitly represent "unspoken" guidelines or understandings, such as who should sign up for the next patient, or which consultants may require additional prompting to see an ED patient. However, information presented for other purposes (available staff resources vs. patient workload; typical wait times for consultants) might provide implicit support for navigating these challenges. Choices made to support, or not, purposes at the higher levels of the hierarchy translate to nodes that were included at the lower levels (e.g., processes regarding caring for the patient, maintenance of situation awareness [SA], and coordination/communication were maintained, while training and administration were not). Some decisions to include information, such as about specific facilities and equipment but not supplies, were due to the available project resources and time.

Conclusions

This research provides an example of how information collected through a work domain analysis can be utilized in an iterative design process to design displays for a novel EDIS. Using CE methods, the research team was able to systematically identify the information needs and goals of the ED and use that information to create display areas and visualizations, which supported the ED purposes of providing quality care and serving as a gatekeeper to the hospital. Other purposes (i.e., education, financial management) were not integrated into these displays but could be similarly developed to support hospital personnel other than ED caregivers. A number of novel information needs were identified, including some that require additional research to create reliable measurement techniques.

Appendix A

Full screen shots of three display areas (see Figures 4.11 through 4.13).

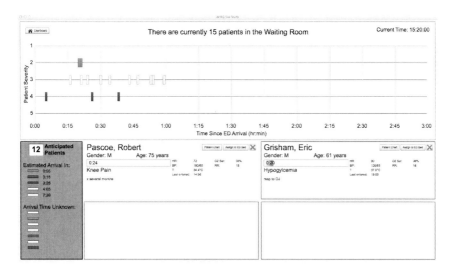

FIGURE 4.11
Complete waiting room display. (Courtesy of Ann Bisantz, University at Buffalo, The State University of New York.)

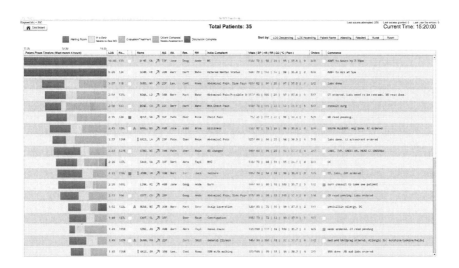

FIGURE 4.12
Complete patient progress display. (Courtesy of Ann Bisantz, University at Buffalo, The State University of New York.)

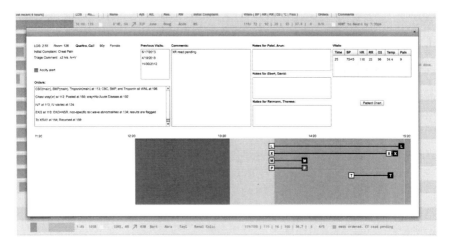

FIGURE 4.13
Detailed patient progress display, accessed by clicking on a specific patient in the complete patient progress display. (Courtesy of Ann Bisantz, University at Buffalo, The State University of New York.)

Acknowledgments

The project described was supported by Grant number R18HS020433 from the Agency for Healthcare Research and Quality (AHRQ). Its contents are solely the responsibility of the authors and do not necessarily represent the official views of AHRQ. The authors acknowledge the research team members and subject matter experts Robert Wears and Shawna Perry; research team members Li Lin, Robert Stephens, Sabrina Cascucci, Longshen Sun, Angelica Hernandez, Natalie Benda, and Vicki Lewis; and subject matter expert Jessica Castner for their various contributions throughout the modeling, interface conceptualization, display implementation, and usability evaluation phases of this project.

References

Ashoori, M., and Burns, C. M. (2013). Team cognitive work analysis: Structure and tasks. *Journal of Cognitive Engineering and Decision Making, 7,* 123–140.
Ashoori, M., Burns, C. M., Momtahan, K., and d'Entremont, B. (2014). Using team cognitive work analysis to reveal healthcare team interactions in a labour and delivery unit. *Ergonomics, 57*(7), 973–986.
Aspden, P., Corrigan, J. M., Wolcott, J., and Erickson, S. M. (2004). *Patient Safety: Achieving a New Standard for Care.* Washington, DC: National Academies Press.

Berg, M. (2003). *Health Information Management: Integrating Information and Communication Technology in Health Care Work.* London: Routledge.

Bisantz, A. M., Pennathur, P. R., Guarrera, T. K., Fairbanks, R. J., Perry, S. J., Zwemer, F., and Wears, R. L. (2010). Emergency department status boards: A case study in information systems transition. *Journal of Cognitive Engineering and Decision Making, 4*(1), 39–68.

Bisantz, A., and Roth, E. (2007). Analysis of cognitive work. *Reviews of Human Factors and Ergonomics, 3*(1), 1–43.

Bisantz, A. M., Roth, E., Brickman, B., Gosbee, L. L., Hettinger, L., and McKinney, J. (2003). Integrating cognitive analyses in a large-scale system design process. *International Journal of Human-Computer Studies, 58*(2), 177–206.

Bisantz, A. M., and Vicente, K. J. (1994). Making the abstraction hierarchy concrete. *International Journal of Human-Computer Studies, 40*(1), 83–117.

Brennan, T. A., Leape, L. L., Laird, N. M., Hebert, L., Localio, A. R., Lawthers, A. G., Newhouse, J., Weiler, P., and Hiatt, H. H. (1991). Incidence of adverse events and negligence in hospitalized patients. *New England Journal of Medicine, 324*(6), 370–376.

Burns, C. M., and Bisantz, A. M. (2008). Advances in the application of cognitive work analysis. In Bisantz, A. M., and Burns, C. M. (Eds), *Applications of Cognitive Work Analysis* (pp. 1–14). Boca Raton, FL: CRC Press.

Burns, C. M., Bisantz, A. M., and Roth, E. M. (2004). Lessons from a comparison of work domain models: Representational choices and their implications. *Human Factors, 46*(4), 711–727.

Burns, C. M., Bryant, D. J., and Chalmers, B. A. (2000). A work domain model to support shipboard command and control. Paper presented at the *Systems, Man, and Cybernetics, 2000 IEEE International Conference.*

Burns, C. M., Enomoto, Y., and Momtahan, K. (2008). A cognitive work analysis of cardiac care nurses performing teletriage. In Bisantz, A., and Burns, C. M. (Eds.), *Applications of Cognitive Work Analysis* (pp. 149–174). Mahwah, NJ: Lawrence Erlbaum and Associates.

Burns, C. M., and Hajdukiewicz, J. (2004). *Ecological Interface Design.* Boca Raton, FL: CRC Press.

Chisholm, C. D., Collison, E. K., Nelson, D. R., and Cordell, W. H. (2000). Emergency department workplace interruptions are emergency physicians "interrupt-driven" and "multitasking"? *Academic Emergency Medicine, 7*(11), 1239–1243.

Clark, L. N., Guarrera, T. K., McGeorge, N. M., Hettinger, A. Z., Hernandez, A., LaVergne, D. T., Benda, N., Perry, S., Wears, R., Fairbanks, R., and Bisantz, A. M. (2014). Usability evaluation and assessment of a novel emergency department IT system developed using a cognitive systems engineering approach. *Proceedings of the 2014 Symposium on Human Factors and Ergonomics in Healthcare.*

Edraki, A., and Milgram, P. (2004). Designing an information querying interface for a rheumatoid arthritis patient record system. *Proceedings of the Human Factors and Ergonomics Society 48th Annual Meeting, 48,* 1634–1637.

Effken, J. A., Brewer, B. B., Logue, M. D., Gephart, S. M., and Verran, J. A. (2011). Using cognitive work analysis to fit decision support tools to nurse managers' work flow. *International Journal of Medical Informatics, 80,* 698–707.

Fairbanks, R. J., Karn, K. S., Caplan, S. H., Guarrera, T. K., Shah, M. N., and Wears, R. L. (2008). Use error hazards from a popular emergency department information system. Paper presented at the *Usability Professionals Association 2008 International Conference.* Baltimore.

Hajdukiewicz, J. R., Doyle, D. J., Milgram, P., Vicente, K. J., and Burns, C. M. (1998). A work domain analysis of patient monitoring in the operating room. *Proceedings of the Human Factors and Ergonomics Society Annual Meeting, 42*(14), 1038–1042.

Institute of Medicine. (2006). *Hospital-Based Emergency Care at the Breaking Point.* Washington, DC: National Academy Press.

Institute of Medicine Committee on Quality of Health Care in America. (2001). *Crossing the Quality Chasm: A New Health System for the 21st Century.* Washington, DC: National Academies Press.

Jiancaro, T., Jamieson, G. A., and Mihailidis, A. (2013). Twenty years of cognitive work analysis in health care a scoping review. *Journal of Cognitive Engineering and Decision Making, 8*(1), 3–22.

Kohn, L. T., Corrigan, J. M., and Donaldson, M. S. (1999). *To Err Is Human: Building a Safer Health System* (Vol. 627). Washington, DC: National Academies Press.

Littlejohns, P., Wyatt, J. C., and Garvican, L. (2003). Evaluating computerised health information systems: Hard lessons still to be learnt. *BMJ: British Medical Journal, 326*(7394), 860.

Lopez, K. D., Gerling, G. J., Cary, M. P., and Kanak, M. F. (2010). Cognitive work analysis to evaluate the problem of patient falls in an inpatient setting. *Journal of the American Medical Informatics Association, 17*, 313–321.

Miller, A., Scheinkestel, C., and Steele, C. (2009). The effects of clinical information presentation on physicians' and nurses' decision-making in ICUs. *Applied Ergonomics, 40*, 753–761.

Rasmussen, J., Pejterson, A., and Goodstein, L. (1994). *Cognitive Systems Engineering.* New York: Wiley and Sons.

Rogers, M. L., Patterson, E. S., Woods, D. D., and Render, M. L. (2006). Cognitive work analysis in health care. In Carayon, P. (Ed.), *Handbook of Human Factors and Ergonomics in Health Care* (pp. 465–473). Mahwah, NJ: Lawrence Erlbaum.

Roth, E. M., and Bisantz, A. M. (2013). Cognitive work analysis. In Lee, J. D., and Kirlik, A. (Eds.), *The Oxford Handbook of Cognitive Engineering* (pp. 240–260). New York: Oxford University Press.

Sharp, T. D., and Helmicki, A. J. (1998). The application of the ecological interface design approach to neonatal intensive care medicine. *Proceedings of the Human Factors and Ergonomics Society Annual Meeting.*

Stead, W. W., and Lin, H. S. (2009). *Computational Technology for Effective Health Care: Immediate Steps and Strategic Directions.* Washington, DC: National Academies Press.

Truxler, R., Roth, E., Scott, R., Smith, S., and Wampler, J. (2012). Designing collaborative automated planners for agile adaptation to dynamic change. *Proceedings of the Human Factors and Ergonomics Society Annual Meeting, 56*(1), 223–227.

Vernon, D., Reising, C., and Sanderson, P. M. (2002). Ecological interface design for Pasteurizer II: A process description of semantic mapping. *Human Factors, 44*(2), 222–247.

Vicente, K. J. (1999). *Cognitive Work Analysis: Toward Safe, Productive, and Healthy Computer-based Work.* Mahwah, NJ: Lawrence Erlbaum Associates.

5

Displays for Health Care Teams: A Conceptual Framework and Design Methodology

Avi Parush

CONTENTS

Introduction

The design and use of displays can be an effective approach to support team-work in health care. The main objective of the chapter is to offer a practical cognitive engineering approach to designing team displays. This chapter reviews briefly the typical challenges to effective teamwork in health care, and the typical uses of displays in health care, and then offers a conceptual framework for team displays. A practical two-phase methodology to design team displays in health care will be outlined in detail.

Challenges to Effective Teamwork in Health Care

In most health care contexts, teams are composed of professionals from sev-eral medical, nursing, and allied health professions. Each of the professions typically trains separately and differently, and they do not necessarily work with each other regularly (Henderson et al. 2007; Kramer et al. 2010a,b). They have different practices and procedures and are responsible for differ-ent aspects and phases of the care process. In addition, team composition is often ad hoc. Very often, health care workers who do not know each other, or do not work with each other regularly, end up as a team due to unforeseen circumstances. And in some cases, it is difficult to discern roles, as in the case for example of trauma team resuscitations, where all team members arrive dressed in identical gowns, masks, and hats.

Different clinical situations with different interprofessional compositions of health care teams result in many team structure permutations. Teams are often hierarchical in nature, with one health care professional serving as the leader. Physicians typically take the leadership role, and a "power distance" is often created (Leonard et al. 2004; Alvarez and Coiera 2006; O'Byrne et al. 2008).

Health care workers in various contexts typically require large amounts of information to conduct their duties. When overburdened, this information load not only can hinder effectiveness but also can harm patients (Imhoff et al. 2001; Salas et al. 2007; Meyfroidt 2009). In addition, the information required by health care workers often arrives from multiple and very diverse sources.

The various characteristics of teamwork in health care described above are often associated with communication breakdowns and information loss. Studies have shown that large portions of in-hospital deaths, adverse events, accidents, and errors were associated with communication failures (Wilson et al. 1995; Bhasale et al. 1998; Cooper et al. 2002; Gawande et al. 2003). In addition, organizational structures of hospitals, surgical teams (Edmondson 2004), and communication failures (e.g., Coiera and Tombs 1998; Coiera 2000; Lingard et al. 2004; Alvarez and Coiera 2006; Parush et al. 2008, 2012) were associated with increased error rates and adverse events.

The characteristics of health care teams were also associated with communication failures across professions, particularly in patient care handoffs, and among the reasons cited were the boundaries created by different "social worlds" (Horwitz et al. 2009). A by-product of the power distance is a limited extent of communication between nurses and physicians directly (Donchin et al. 1995). A large and compelling body of research shows that communication failures are a significant problem in health care teamwork.

Displays to Support Team Performance

There is evidence that effective teamwork is critical to safety and quality of care in health care. There is also evidence that effective teamwork can break down under various circumstances. Those circumstances are often unavoidable so there is a need to look for more tools to support team performance. One such approach is the use of augmentative and supportive technology. The use of intelligent displays is one of the solutions proposed to support teamwork in various domains (Roth et al. 1998; McKinsey and Company 2002; Salas et al. 2008; Gouin and Vernik 2010; Parush et al. 2011).

Display Devices in Support of Teamwork

Other Domains

Display devices supporting information sharing within teams have been developed and evaluated in various domains (Bolstad and Endsley 1999, 2000; Gorman et al. 2006; Kulyk et al. 2008). Several systems have been designed for the purposes of supporting work tasks or collaborative work.

Display devices to support collaborative activities and teamwork have been studied in domains such as military command-and-control (Jedrysik et al. 2000; Wark et al. 2004; Scott et al. 2007), aviation and air traffic control (Bolstad and Endsley 1999; Dudfield et al. 2001; Darling and Means 2004), and naval ship command (Jenkin 2004; Dominguez et al. 2006) and even urban firefighting (Jiang et al. 2004).

In Health Care

The use of display devices to support collaboration and teamwork in health care will be discussed in terms of the functions they serve: (1) management and administration, (2) guidance and decision-making support, and (3) information sharing.

Management and Administration

Information about personnel, shift work, workload distribution, and status of various cases, treatment rooms, etc., is shared among health care personnel. These can appear on various artifacts (e.g., Xiao et al. 2001). Team-oriented technology can be useful for this purpose as well. The use of an electronic whiteboard displaying management and administrative information has been studied in various health care contexts (France et al. 2005; Wears and Perry 2007; Aronsky et al. 2008; Bossen and Jensen 2008).

Guidance and Decision-Making Support

Health care workers often follow procedures, protocols, and algorithms. Job aids such as checklists and care algorithms are some examples. Team-oriented displays can be used to provide such structured guidance to their tasks (Porter et al. 2004; Patel et al. 2008; Fitzgerald et al. 2011).

Information Sharing

Team members share information among themselves as an inherent part of teamwork processes. One of the important aspects of display technologies in supporting teamwork is by serving as a channel for sharing information and providing a comprehensive and timely situation picture (Levine et al. 2005; Lai et al. 2006; Kramer et al. 2008; Sarcevic et al. 2010, 2012; Parush et al. 2011; Pennathur et al. 2011). Using information-sharing displays as part of teamwork is critical since many of the failures in health care teamwork are associated with communication breakdowns, problems with information sharing, and information loss.

Conceptual Framework for Considering Information-Sharing Displays

The existence of information displays designed to support teamwork has been well described. However, how does one consider whether displays are a relevant solution for a given context? Authors have proposed general frameworks (e.g., Douglas et al. 2007). This chapter offers a conceptual framework specifically oriented to health care contexts. The framework consists of two key aspects: (1) context and needs and (2) possible solutions.

Context and Needs

Among the factors that influence information sharing within health care teams are the number of information sources and the need for sharing that

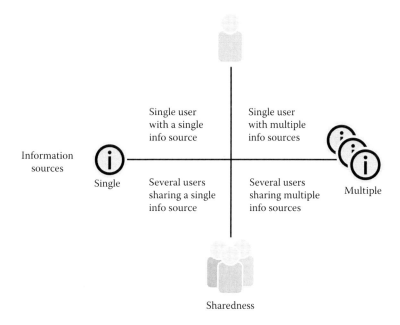

FIGURE 5.1
Two dimensions expressing the context and needs in health care teams: the number of information sources and the need to share those information sources.

information among several health care professionals. The contextual factors introduced by the combination of information sources and the number of users of those data are depicted schematically in Figure 5.1.

Number of Information Sources

Health care professionals utilize information that can originate from a single, several, or multiple sources of data. The challenge and complexity of providing and presenting the information increase as the number of information sources increases.

Level of Sharedness

A single health care professional role can utilize information originating from a single source or multiple sources. Or several or all members of the health care team can utilize information originating from a single source or multiple sources. The level of information sharedness increases when more than one health care professional utilizes information from common sources. The challenge and complexity of presenting the information increase as the level of sharedness among team members increases. As was indicated earlier, most health care contexts are characterized by all team members requiring information from multiple sources. Thus, the need to find effective solutions for displaying information from multiple sources to multiple health care workers is a common challenge.

Possible Solutions

The combination of multiple information sources and the need of several team members to utilize the information introduce challenges to supporting team information sharing. The challenge is in how to process the information, particularly from multiple sources, and how to display it effectively to different team members. The possible solutions resulting from the combination of different ways to process the information and the manner of displaying it are depicted schematically in Figure 5.2.

Information Processing and Presentation

Information originating from a single source can be visualized in various manners without the need to combine or integrate it with any other information. For example, prothrombin time (PT, also expressed as INR), a vital clinical parameter reflecting the level of blood thinning, can be simply displayed as a one-time value. However, often there is a need to utilize information originating from several sources, such as past values, medication administered (particularly warfarin, a common medication used to increase the PT/INR), and other medical interventions such as the administration of reversal agents. In this case, visualizing the information in an integrated fashion can support health care professionals in effectively receiving and utilizing the information.

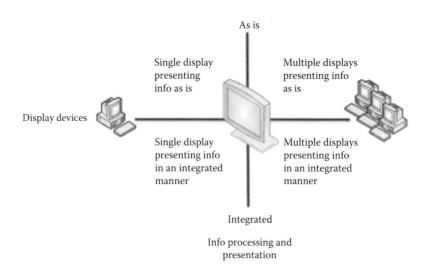

FIGURE 5.2
Two dimensions expressing ways that information can be processed for display and the number and size of screens on which it can be displayed.

There can be several ways to present information from multiple sources. One possibility is to have the information from the various sources visualized and only colocated in a spatial arrangement (e.g., in separate tiles or windows or screens). This is less desirable as it fails to take advantage of opportunities to integrate the information to aid in decision making. Another possibility is to integrate the information from the difference sources and to synthesize or abstract it into a new composition of the information that meets the cognitive needs of the users. This requires an understanding of the cognitive work of the user that allows anticipation of their data needs (the right data at the right time, displayed in the right context). Ecological interface design (EID) is an approach for integrating and abstracting multiple-source information in a way that meets the cognitive needs of the user (e.g., Burns and Hajdukiewicz 2004; see also Bolstad and Endsley 2000 on abstracted shared displays). An integrated display for the operating room (OR) was developed and reported by several researchers (Levine et al. 2005; Lai et al. 2006; Parush et al. 2011). This work emphasizes the integration of all critical information into a single large display to enhance situation awareness. Moreover, the displays are intended to facilitate team orientation during the procedure and during shift changes.

The challenge in integrating and synthesizing multiple-source information for teams is developing an understanding of what critical information requires sharing among team members. The latter part of this chapter will present a methodology for meeting this design challenge.

The processing and presentation of the information from a single source or multiple sources is one aspect of the challenge in supporting information sharing within teams. The other aspect of the challenge is how to display physically the information in an effective manner.

Information-Displaying Devices

Information from single or multiple sources can be displayed on a single display device. The size of that single display device can range between small and large. The screen size of choice depends on contextual factors such as the amount of data to be displayed, the number of people receiving information from the display, and whether one is near or far from the display while executing additional concurrent tasks. For example, an anesthetist in the OR can be very close to the display of the anesthesia records, and if the amount of data is small, the device need not be large. However, the surgeon in the OR cannot be too close to the device presenting patient vital signs, and thus, the display device choice would be a larger screen that can be viewed from a distance.

Information can also be displayed on multiple display devices. Single source information would be displayed on multiple devices when it is shared by several team members, and each requires the information at a different location (e.g., in a large OR spread in a hospital unit), even when they are all colocated. In that case, information could be cloned across the multiple displays, or it might employ different visualizations to meet the needs of the different user groups. However, in some health care contexts with colocated teams, the solution of multiple displays is not effective. Similarly, multiple-source information is sometimes displayed on multiple systems, requiring users to toggle between programs, often with different logons. Such a solution can have a strong negative effect on efficiency, particularly in the high-workload environments found in most health care contexts. The presentation of information for collaboration, coordination, and information sharing, from either a single source or multiple sources, on a single or several large screen displays seems to be a more effective solution (Smith and Duggar 1965; Roth et al. 1998; Dudfield et al. 2001; Izadi et al. 2005; Gouin and Vernik 2010; Wallace et al. 2011).

Summary and Transition

The conceptual framework outlined in this section can guide initial needs assessment and solution exploration with respect to the use of team displays in health care. Needs are identified in terms of the type and the scope of information required by the various team members and the extent to which that information needs to be shared by some or all team members. The methodology proposed here is aimed at formulating requirements and designing the visual display of information on large screen displays.

Methodology for Designing Displays for Health Care Teams

Methodology Overview

There are various approaches to designing team displays. The methodology presented here is empirical (in contrast to analytic) and includes elements of user-centered design (UCD). It is empirical because it includes collecting real-world data on teams and teamwork in the relevant context. It uses UCD elements because end users are involved in various phases of the research, design, and development. In addition, an iterative approach is a characteristic of the UCD process whereby there can be several iterations until the desired solution is achieved. Chapter 4 of this book also presented an iterative method based on both cognitive engineering and UCD for designing a display for a health care team.

The methodology is composed of two iterative phases: research and design. The main objective of the research phase of the methodology is to identify the information instances and categories shared by all or at least several team members. The research phase consists of (1) data collection, (2) analysis, and (3) synthesis (Figure 5.3). The research outcomes are translated

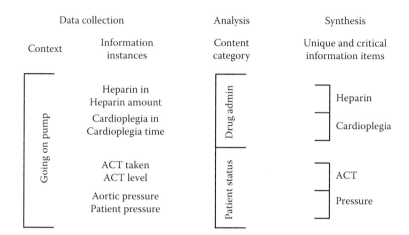

FIGURE 5.3
Examples of shared information in cardiac surgery for the three research steps: data collection, analysis, and synthesis.

into display design requirements, which are then implemented and tested. The design phase consists of (1) conceptual design, (2) detailed design, and (3) prototype and evaluation.

Cardiac OR as a Case Study

The approach to designing team displays for health care contexts is illustrated here with a case study of a display for the cardiac OR team. Cardiac surgery involves the surgical treatment of the heart and/or main blood vessels leading to and from the heart. Each procedure included a team composed of at least seven health care workers: an attending surgeon and a surgery resident, an anesthesiologist and respiratory therapist, perfusionist, scrub nurse, and circulating nurse. The work typically follows a well-known, well-practiced sequence that varies very little from one instance to another. Team members typically know each other, and operations are planned and scheduled.

Methodological Phase 1: Research

Step 1: Data Collection

The key objective of the research component is to gain understanding of teams' workflow, coordination, and collaboration patterns and have a particular focus on the information required and used by team members. The research approach is qualitative in nature and includes two key data collection activities: (1) interviews and (2) observational studies.

Interviews

The research started with a series of semistructured interviews with representatives of medical and nursing professions participating in cardiac surgery (two surgeons, two anesthetists, two perfusionists, and three nurses). The purpose of the interviews was to provide detailed orientation and explanations of the open-heart surgery procedure.

Observations

The subsequent research phase consisted of observational studies. Sixteen open-heart surgeries, all including a heart–lung bypass machine (i.e., the heart is bypassed, and an external pump circulates blood to the body), were observed in a medium-sized institute specializing in cardiology and cardiac surgery. Overall, the observations involved four different surgeons, four different surgery residents, five pump operators (perfusionists or respiratory therapist), five different anesthetists, and scrub and circulating nurses totaling 15 different nurses. Two observers observed participants during all communicative interactions (e.g., surgeon talking to the scrub nurse) and captured independently the information conveyed (e.g., surgeon asking the perfusionist if the ACT was done).

Step 2: Analysis

As was mentioned earlier, the research focus was on the shared information in order to utilize it later for the team display design. To achieve this objective, the interviews and observations were analyzed separately, and then the key findings from each were merged in order to provide a common basis for synthesizing the display requirements.

Interviews

The theoretical underpinning of the interview analysis is the assumption that team members share portions of their mental models about their tasks and teamwork. The analysis approach strives to find the commonalities across all professions and team members as opposed to looking for differences and gaps. Specifically, the analysis of the interviews focused on two aspects: (1) event workflow and (2) information categories. The first aspect, event workflow, consists of the sequence of activities and interactions and various activities taking place in parallel. Data in this aspect were assumed to reflect what the various health care professionals think about how they are supposed to work with each other and throughout the event. The second aspect, information categories, consists of the various clinical and nonclinical topics the team members cover or deal with during the event. Data in this aspect were assumed to reflect how team members chunk and organize the information in a way that they can understand it or convey it effectively to each other.

In order to benefit from the analysis in designing team displays, the primary objective of the interview analysis is to identify the shared information instances. In order to meet this objective, the event workflow was depicted as a flow chart resembling the operational sequence diagram or what is more recently referred to in the software engineering domain as swim lanes. This visualization approach supports identifying portions in the overall work-flow consisting of interactions between team members and the information they share. In addition, it provides the timing of interactions and informa-tion sharing within the overall workflow. This can later be used in determin-ing the dynamics of information presentation on the team display.

The second part of the analysis is grouping all information instances into information categories. These are summarized in a table outlining the infor-mation categories along with an explanation and an example. Common cat-egories include patient history, patient status, drug administration, medical actions, the timing of various actions, and medical equipment, among others.

Observations

Communication analysis was employed on the observational data collected in the case study. Similar to the interview analysis, the primary objective was to uncover information shared among team members and the timing of the exchange. The communication analysis methodology is based on human communication theories (e.g., Schramm 1954) and the conversation analy-sis framework (e.g., Schegloff 1987; Pomerantz and Fehr 1997). The approach involves segmenting the verbal communications in a given context into meaningful sequences that serve a communicative purpose (e.g., a question, a directive, a confirmation, etc., about a specific topic). Such basic elements in analyzing human communication appear in diverse domains (e.g., Heath and Luff 1991; Orasanu 1994; Bowers et al. 1998; Davies 2005; Nemeth et al. 2005). Information categories were then identified in each exchange. In a similar fashion to the interview analysis, the information categories were outlined in tables along with an explanation and an example. As was the case with the interview analysis, similar information categories were found in the communication analyses of both cases and include categories such as patient history, patient status, drug administration, medical actions, the tim-ing of various actions, and medical equipment (Parush et al. 2009, 2011, 2012).

Step 3: Synthesis

The outcome of the analysis was the mapping of many shared information instances into information categories as a function of the context in which they appear (i.e., the timing of their appearance in the event flow). This resulting mapping still contained a very large number of information items, too large to be considered for a display design. The working assumption in designing the team display is that the team members do not require all the information presented to them all the time. Thus, the objective of the subse-quent synthesis step is evaluate the information required by the team and the timing at which they use the information.

Scoping was done by interviewing relevant health care professionals and asking them to rate the essentiality and criticality of information items. Specifically, each information item was rated by three questions, as a function of the event flow. The first question was, "How important is it to share this information with members of the team so that they can provide a backup for the person responsible for this information?" with a rating scale from 1 to 7 (with 1 being very important and 7 not important at all). The second question was, "How important is it to share this information with members of the team so that they can do *their* job better?" with the same rating scale. Finally, respondents were asked to indicate who among the team members requires that information item.

The analysis of all the responses resulted in a list of unique information items. These unique information items kept their mapping to information categories and the specific context in the event flow. The unique items were assigned into three priority groups as a function of the phases in the event flow. The highest priority group included items with a broad consensus on the very high criticality of sharing them among all team members. The medium priority included all items with a mean of 2–4 on the scales. Finally, the third priority included items with a broad consensus that they need not be shared with all the team members. The working assumption was that the list of the highest priority items constituted the requirements of what needs to be displayed to the team.

Methodological Phase 2: Design

Step 1: Conceptual Design

The high-priority information items and the time they are utilized during the surgery workflow formed the foundation for conceptual design. The challenge in conceptual design is to determine the appropriate grouping criteria and their trade-offs with respect to achieving the goal of presenting to the team the right information at the right time. Once such groups are determined, the process can proceed into the detailed design of the visual presentation.

In the cardiac surgery case, four grouping criteria were considered based on the information categories uncovered in the research phase and in collaboration with the relevant health care professionals. The four grouping criteria and the strengths and weaknesses of each are outlined in Table 5.1.

By consideration of the trade-offs between the strengths and weaknesses of the different criteria, and in consultation with the relevant health care professionals, a composite grouping criterion was adopted. The composite grouping included the physical and/or anatomical grouping along with functional groups. There was a distinction between static and dynamic information within each of these groups.

The conceptual design consisted of three main groups: (1) static patient information and dynamic patient status information; (2) anatomical/physical components—the patient's heart and lungs and the pump; and (3) the

TABLE 5.1

Strengths and Weaknesses Comparison of the Four Grouping Criteria for the
Conceptual Design of the Cardiac Team Display

Conceptual Grouping Criteria	Strengths	Weaknesses
Static information items (e.g., patient information) vs. dynamic parameters (e.g., vital signs)	Strong visual guidance for information searching and monitoring	Weak indication of causal relations between parameters (e.g., impact of drugs on BP level)
Functional groups (e.g., drug administration parameters)	A strong visual affinity between information items related to similar functions	Lack of context sensitivity to the event flow Weak link to physical and anatomical components (e.g., the pump, lungs, etc.)
Event flow (e.g., patient prep parameters, and then going on pump, etc.)	A strong workflow-related information	Can be too dynamic, thus imposing a greater challenge to visual search and reading
Physical and/or anatomical grouping (e.g., heart, lungs, drugs, machine, etc.)	A strong visual affinity between various critical components of the procedure	Lacks the sensitivity to the procedural phase and link to functional groups such as drugs or patient status

composite group, the functions that link between the anatomical and functional components (drugs and other clinical actions). The presence or absence of relevant information items in each group was driven by the event flow. For example, the information about certain drugs or tests was presented only during the relevant surgery phase and not all the time. The resulting conceptual design is a wireframe depicted as in Figure 5.4.

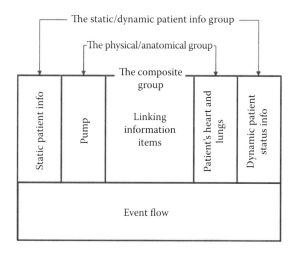

FIGURE 5.4
High-level wireframe sketch of the conceptual design of the cardiac team display.

Step 2: Detailed Design

The first step in the detailed design was to determine the visual language aspects and graphically design each element. In the cases described here, the decision was to have a realistic graphical representation of each element as opposed to an abstract form. Each information item was designed to represent visually and realistically the content it conveys. Figure 5.5 shows a few examples of the detailed visual design of several information items in the cardiac surgery case.

The second step was to visually lay out all groups and elements. The visual concept for the display followed two key design principles: (1) proximity—closely related information items, according to a given grouping criterion (e.g., administering the drug heparin and the ACT test showing the blood-thinning level), were displayed with a physical proximity to clearly convey

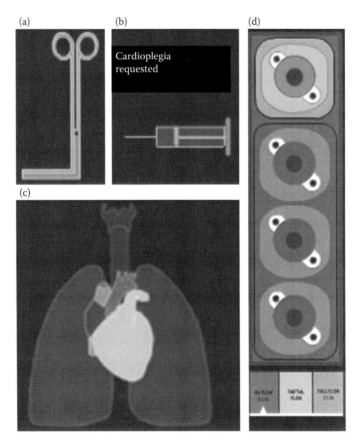

FIGURE 5.5
Graphic examples of several information items in the cardiac team display. (a) Cross clamps; (b) the pump; (c) heart and lungs; and (d) drug-related action.

their relations; and (2) redundancy—critical information items can be displayed in more than one fashion or in more than one place to ensure that the information is not missed.

In the cardiac case, the representation of the patient's lungs and heart, on one hand, and the heart–lung machine, on the other hand, were visually placed separately, similar to how they are placed in the OR. A representation of the cannulas (tubes) when connected to the patient and the machine was used to create a process flow representation (e.g., on or off pump). Finally, the drugs and the ACT levels, which are parameters that help manage the human–machine flow, were placed in between these two elements. With respect to redundancy, all critical parameters and actions were presented in at least two places. For example, the indication on heparin administration was depicted both in the composite group area next to the resulting ACT and in the event flow representation.

Once all information items, at their various states, were fully designed, the entire display was configured according to the conceptual design wireframe. The final detailed design, based on the conceptual design, is shown in Figure 5.6.

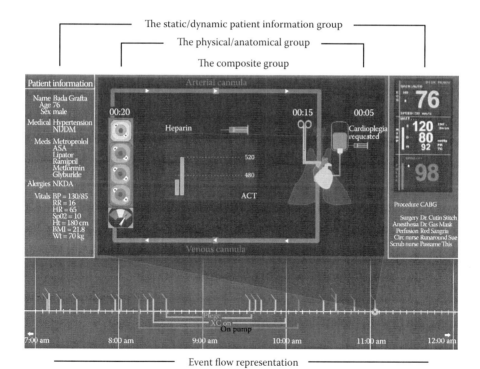

FIGURE 5.6
Detailed visual design of the cardiac team display and links to the underlying conceptual groups.

Step 3: Prototype and Evaluation

An important characteristic of UCD is the iterative nature of the process. Early evaluation of the design concepts and details can help detect and identify problems, and these may lead to additional research and design iterations. Early evaluation requires a quick implementation of the design, typically achieved by developing a low- or high-fidelity prototype of the design.

In the cases described here, the visual design was implemented as a dynamic display, and the behavior of the displayed information (such as the appearance or disappearance of parameters, the value of parameters, etc.) was remotely and manually controlled by the evaluation administrator. The collaborating health care professionals designed evaluation scenarios, and these were simulated following the "Wizard of Oz" approach. In other words, the evaluating health care professional observed the screen, the evaluation administrator "drove" the display following the scenario, and the evaluation participant provided comments and responded to questions assessing whether the information was conveyed effectively. Results indicated that the participants understood the information and the situation based on the display (Parush et al. 2011).

Research Agenda and Future Directions

This chapter presented a cognitive engineering approach to designing displays for a specialized cardiac surgery health care team. Based on researching aspects in team cognition, primarily via team communication analysis, the methodology described here covers designing and developing displays to present multisource information required by health care teams. The success of team displays in various domains, particularly those implemented as large screen displays, is encouraging and provides a valid basis for pursuing similar directions in health care.

More research should provide empirical evidence on the effectiveness of team displays in health care contexts. Such research should include realistic scenarios in full-scale high-fidelity simulators in which the use of the display is tested. Studies should look into the impact of the display on both clinical/technical aspects and nontechnical teamwork. Specifically, studies should assess implications of having the display on clinical performance and outcomes, care accuracy and quality, mitigation of harm, and improvement of patient safety. In addition, research should assess whether and how the presence of team displays influences teamwork processes.

There are technological challenges that must be considered, including acquiring data from multiple sources in real time, processing and integrating the information, and driving the display in real time. Thus, aside from assessing the effectiveness of the display in supporting teamwork, much work is still needed to develop and test the technological aspects of such displays.

References

Alvarez, G., and Coiera, E. (2006). Interdisciplinary communication: An uncharted source of medical error? *Journal of Critical Care*, 21, 236–242.

Aronsky, D., Jones, I., Lanaghan, K., and Slovis, C.M. (2008). Supporting patient care in the emergency department with a computerized whiteboard system. *Journal of the American Medical Informatics Association*, 15(2), 184–194.

Bhasale, A.L., Miller, G.C., Reid, S.E., and Britt, H.C. (1998). Analysing potential harm in Australian general practice: An incident-monitoring study. *The Medical Journal of Australia*, 169, 73–76.

Bolstad, C.A., and Endsley, M.R. (1999). Shared mental models and shared displays: An empirical evaluation of team performance. *Proceedings of the 43rd Annual Meeting of the Human Factors & Ergonomics Society*, 213–217.

Bolstad, C.A., and Endsley, M.R. (2000). The effect of task load and shared displays on team situation awareness. *Proceedings of the 14th Triennial Congress of the International Ergonomics Association and the 44th Meeting of the Human Factors and Ergonomics Society*, 189–192.

Bossen, C., and Jensen, L. (2008). Implications of shared interactive displays for work at a surgery ward: Coordination, articulation work and context-awareness. *The 21st IEEE International Symposium on Computer Based Medical Systems*, 464–469.

Bowers, C.A., Jentsch, F., Salas, E., and Braun, C.C. (1998). Analyzing communication sequences for team training needs assessment. *Human Factors*, 40, 672–679.

Burns, C.M., and Hajdukiewicz, J. (2004). *Ecological Interface Design*. Boca Raton, FL: CRC Press.

Coiera, E., and Tombs, V. (1998). Communication behaviours in a hospital setting: An observational study. *The British Medical Journal*, 316, 673–676.

Coiera, E. (2000). When conversation is better than computation. *Journal of the American Medical Informatics Association*, 7, 277–286.

Cooper, J.B., Newbower, R.S., Long, C.D., and McPeek, B. (2002). Preventable anesthesia mishaps: A study of human factors. *Quality and Safety in Healthcare*, 11, 277–282.

Darling, E., and Means, D.C. (2004). A methodology for unobtrusively determining the usage of C2 data walls. *The 10th International Command and Control Research and Technology Symposium: The Future of C2*.

Davies, J.M. (2005). Team communication in the operating room. *Acta Anaesthesiologica Scandinavica*, 49, 898–901.

Dominguez, C., Long, W.G., Miller, T.E., and Wiggins, S.L. (2006). Design directions for support of submarine commanding officer decision making. *The 2006 Undersea HSI Symposium*, Mystic, CT.

Donchin, Y., Gopher, D., Olin, M., Badihi, Y., Biesky, M., and Sprung, C.L. (1995). A look into the nature and causes of human errors in the intensive care unit. *Critical Care Medicine*, 23, 294–300.

Douglas, L., Alvera, D., and Havig, P. (2007). Shared displays: An overview of perceptual and cognitive issues. *Proceedings of the Twelfth International Command and Control Research and Technology Symposium,* June 19–21 2007, Newport, RI.

Dudfield, H.J., Macklin, C., Fearnley, R., Simpson, A., and Hall, P. (2001). Big is better? *Human Interfaces in Control Rooms, Cockpits and Command Centers,* 48, 304–309.

Edmondson, A.C. (2004). Learning from mistakes is easier said than done. *The Journal of Applied Behavioral Science,* 40, 66–90.

Fitzgerald, M., Cameron, P., Mackenzie, C., Farrow, N., Scicluna, P., Gocentas, R., Bystrzycki, A. et al. (2011). Trauma resuscitation errors and computer-assisted decision support. *Archives of Surgery,* 146, 218–225.

France, D.J., Levin, S., Hemphill, R., Chen, K., Rickard, D., Makowski, R., Jones, I., and Aronsky, D. (2005). Emergency physicians' behaviors and workload in the presence of an electronic whiteboard. *International Journal of Medical Informatics,* 74, 827–837.

Gawande, A.A., Zinner, M.J., Studdert, D.M., and Brennan, T.A. (2003). Analysis of errors reported by surgeons at three teaching hospitals. *Surgery,* 133, 614–621.

Gorman, J.C., Cooke, N.J., and Winner, J.L. (2006). Measuring team situation awareness in decentralized command and control systems. *Ergonomics,* 49, 1312–1325.

Gouin, D., and Vernik, R. (2010). Using large group displays to support intensive team activities in C2. *The 16th International Command and Control Research and Technology Symposium.*

Heath, C., and Luff, P. (1991). Collaborative activity and technological design: Task coordination in London underground control rooms. *Proceedings of the European Conference on Computer Supported Cooperative Work,* 65–80.

Henderson, S., Mills, M., Hobbs, A., Bleakley, A., Boyden, J., and Walsh, L. (2007). Surgical team self-review: Enhancing organizational learning in the Royal Cornwall Hospital Trust. In Cook, M., Noyes, J., and Masakowski, Y. (Eds.), *Decision Making in Complex Environments* (pp. 259–267). Burlington, VT: Ashgate Publishing Company.

Horwitz, L.I., Meredith, T., Schuur, J.D., Shah, N.R., Kulkarni, R., and Grace, Y. (2009). Dropping the baton: A qualitative analysis of failures during the transition from emergency department to inpatient care. *Annals of Emergency Medicine,* 53, 701–710.

Imhoff, M., Webb, A., and Goldschmidt, A. (2001). Health informatics. *Intensive Care Medicine,* 27, 179–186.

Izadi, S., Fitzpatrick, G., Rodden, T., Brignull, H., Rogers, Y., and Lindley, S. (2005). The iterative design and study of a large display for shared and sociable spaces. *Proceedings of the Conference on Designing for User Experience,* 1–59.

Jedrysik, P.A., Stedman, T.A., Moore, J.A., and Sweed, R.H. (2000). Interactive displays for command and control. *Proceedings of the IEEE Aerospace Conference,* 341–351.

Jenkin, C.M. (2004). Team situation awareness: Display technologies in support of maritime domain awareness. White paper available at http://onlinepubs.trb.org/onlinepubs/archive/Conferences/MTS/3C%20JenkinPaper.pdf.

Jiang, X., Hong, J.I., Takayama, L.A., and Landay, J.A. (2004). Ubiquitous computing for firefighters: Field studies and prototypes of large displays for incident command. *Proceedings of the SIGCHI Conference on Human Factors in Computing Systems,* 679–686.

Kramer, C., Foster-Hunt, T., Parush, A., and Momtahan, K. (2008). Design and evaluation of a team-oriented display to facilitate situation awareness and increase patient safety during cardiac surgery. *International System Safety Conference 2008*, August 2008, Vancouver, BC, Canada.

Kramer, C., Parush, A., and Momtahan, K. (2010a). Cross-professional miscommunications and patient safety in thoracic operating rooms. Poster presented to *The Ottawa Hospital Safety Education Day*, November 4, Ottawa, Ontario.

Kramer, C., Parush, A., Brandigampola, S., and Momtahan, K. (2010b). Analysis of cross-professional communication in thoracic operating rooms. In Duffy, V.G. (Ed.), *Advances in Human Factors and Ergonomics in Healthcare* (pp. 307–316). Boca Raton, FL: CRC Press.

Kulyk, O., van der Veer, G., and van Dijk, B. (2008). Situational awareness support to enhance teamwork in collaborative environments. *Proceedings of the 15th European Conference on Cognitive Ergonomics: The Ergonomics of Cool Interaction*, Article No. 5.

Lai, F., Spitz, P., and Brzezinski, P. (2006). Gestalt operating room display design for perioperative team situation awareness. *Studies in Health Technology and Informatics*, 119, 282–284.

Leonard, M., Graham, S., and Bonacum, D. (2004). The human factor: The critical importance of effective teamwork and communication in providing safe care. *Quality and Safety in Health Care*, 13(Suppl 1), i85–i90.

Levine, M., Meyer, M., Brzezinski, P., Robbins, J., Lai, F., Spitz, G., and Sandberg, W. (2005). Usability factors in the organization and display of disparate information sources in the operative environment. *Proceedings of the American Medical Informatics Association Annual Symposium*, 1025.

Lingard, L., Espin, S., Whyte, S., Regehr, G., Baker, G., Reznick, R., Bohnen, J., Orser, B., Doran, D., and Grober, E. (2004). Communication failures in the operating room: Observational classification of recurrent types and effects. *Quality and Safety in Healthcare*, 13, 330–334.

McKinsey and Company. (2002). *McKinsey Report: Increasing FDNY's Preparedness.* New York: McKinsey.

Meyfroidt, G. (2009). How to implement information technology in the operating room and the intensive care unit. *Best Practice & Research Clinical Anaesthesiology*, 23, 1–14.

Nemeth, C.P., Kowalsky, J., Brandwijk, M., O'Connor, M.F., Nunnally, M.E., Klock, P.A., and Cook, R.I. (2005). Distributed cognition: How hand-off communication actually works. *Anesthesiology*, 103, A1289.

O'Byrne, W.T., Weavind, L., and Selby, J. (2008). The science and economics of improving clinical communication. *Anesthesiology Clinics*, 26, 729–744.

Orasanu, J. (1994). Shared problem models and flight crew performance. In Johnston, N., McDonald, N., and Fuller, R. (Eds.), *Aviation Psychology in Practice* (pp. 255–285). Aldershot, UK: Ashgate.

Parush, A., Momtahan, K., Foster-Hunt, T., Kramer, C., Holm, C., and Nathan, H. (2008). Critical information flow analysis in the cardiac operating room. Oral presentation at the *International System Safety Conference 2008*, Vancouver, Canada.

Parush, A., Momtahan, K., Foster-Hunt, T., Kramer, C., Hunter, A., and Nathan, H. (2009). A communication analysis methodology for developing cardiac operating room team-oriented displays. *Proceedings of the 53rd Annual Human Factors and Ergonomics Society Conference*, 728–732.

Parush, A., Kramer, C., Foster-Hunt, T., Momtahan, K., Hunter, A., and Sohmer, B. (2011). Communication and team situation awareness in the OR: Implications for augmentative information display. *Journal of Biomedical Informatics*, 44, 477–485. doi: http://dx.doi.org/10.1016/j.jbi.2010.04.002.

Parush, A., Kramer, C., Foster-Hunt, T., McMullan, A., and Momtahan, K. (2012). Exploring similarities and differences in teamwork across diverse healthcare contexts using communication analysis. *Cognition, Technology & Work*, 16, 47–57.

Patel, V.L., Zhang, J., Yoskowitz, N.A., Green, R., and Sayan, O.R. (2008). Translational cognition for decision support in critical care environments: A review. *Journal of Biomedical Informatics*, 41, 413–431.

Pennathur, P.R., Cao, D., Bisantz, A.M., Lin, L., Fairbanks, R.J., Wears, R.L., Perry, S.J., Guarrera, T.K., Brown, J.L., and Sui, Z. (2011). Emergency department patient tracking system evaluation. *International Journal of Industrial Ergonomics*, 41, 360–369.

Pomerantz, A., and Fehr, B.J. (1997). Conversation analysis: An approach to the study of social action as sense making practices. In van Dijk, T.A. (Ed.), *Discourse as Social Interaction* (pp. 64–91). New York: Sage.

Porter, S.C., Cai, Z., Gribbons, W., Goldman, D.A., and Kohane, I.S. (2004). The Asthma Kiosk: A patient centered technology for collaborative decision support in the emergency department. *Journal of the American Medical Informatics Association*, 11, 458–467.

Roth, E.M., Lin, L., Thomas, V.M., Kerch, S., Kenney, S.J., and Sugibayashi, N. (1998). Supporting situation awareness of individuals and teams using group view displays. *Proceedings of the Human Factors and Ergonomics Society 42nd Annual Meeting*, 244–248.

Salas, E., Rosen, M.A., and King, H. (2007). Managing teams managing crises: Principles of teamwork to improve patient safety in the emergency room and beyond. *Theoretical Issues in Ergonomics Science*, 8(5), 381–394.

Salas, E., Cooke, N.J., and Rosen, M.A. (2008). On teams, teamwork, and team performance: Discoveries and developments. *Human Factors*, 50, 540–547.

Sarcevic, A., Marsic, I., and Burd, R.S. (2010). Does size and location of the vital sign monitor matter? A study of two trauma centers. *Proceedings of the American Medical Informatics Association Annual Symposium*, 707–711.

Sarcevic, A., Weibel, N., Hollan, J.D., and Burd, R.S. (2012). A paper-digital interface for information capture and display in time-critical medical work. *Pervasive Computing Technologies for Healthcare*, 17–24.

Schegloff, E.A. (1987). Analyzing single episodes of interaction: An exercise in conversation analysis. *Social Psychology Quarterly*, 50, 101–114.

Schramm, W. (1954). How communication works. In Schramm, W. (Ed.), *The Process and Effects of Mass Communication* (p. 3). Chicago: University of Illinois Press.

Scott, S.D., Wan, J., Rico, A., Furusho, C., and Cummings, M.L. (2007). Aiding team supervision in command and control operations with large-screen displays. *Proceedings of HSIS 2007: ASNE Human Systems Integration Symposium*, March 19–21, 2007, Annapolis, MD.

Smith, S.L., and Duggar, B.C. (1965). Do large displays facilitate group effort? *Human Factors*, 7, 237–244.

Wallace, J.R., Scott, S.D., Lai, E., and Jajalla, D. (2011). Investigating the role of large, shared display in multi-display environments. *Computer Supported Cooperative Work*, 20, 529–561.

Wark, S., Zschorn, A., Broughton, M., and Lambert, D. (2004). FOCAL: A collaborative multimodal multimedia display environment. *Proceedings of SimTect*, 24, 44–78.

Wears, R., and Perry, S.J. (2007). Status boards in accident and emergency departments: Support for shared cognition. *Theoretical Issues in Ergonomics Science*, 8, 371–380.

Wilson, R.M., Runciman, W.B., Gibberd, R.W., Harrison, B.T., Newby, L., and Hamilton, J.D. (1995). The quality in Australian health care study. *Medical Journal of Australia*, 163, 458–471.

Xiao, Y., Lasome, C., Moss, J., Mackenzie, C.F., and Faraj, S. (2001). Cognitive properties of a whiteboard: A case study in a trauma centre. *Proceedings of the Seventh European Conference on Computer-Supported Cooperative Work*, 259–278.

6

Information Modeling for Cognitive Work in a Health Care System

Priyadarshini R. Pennathur

CONTENTS

Information in Health Care

Information and Errors

Inadequate design support for information-based cognitive work continues to cause significant medical errors (Wilson et al. 1995; Kohn et al. 1999; Leape and Berwick 2005; Williams et al. 2007). Problems include availability of and timely access to information, lack of information integration, and poor provisions for sharing and transfer of accurate information (Ash et al. 2004; Brennan et al. 2004; Arora 2005; Calleja et al. 2011). For example, in a landmark study, Leape et al. (1995) found that not having enough information support precipitated 7 of 16 system failures and contributed to 78% of all adverse drug events observed in their study. Similarly, Wilson et al. (1999) found that poor sharing of essential information contributed to over 74% of preventable adverse events. Information discrepancies in medication orders, uncertainties in diagnosis from lack of information, and inadequate clinical decision support also cause medical errors (Barker et al. 2002; Croskerry 2003; Shulman et al. 2005; Koppel et al. 2008; Halbesleben et al. 2010). The potential for expensive medical errors heightens the need to examine how we design, operate, and maintain health care information systems.

Cognitive Work Design Methods and Health Care Information Systems

The philosophical underpinning behind cognitive engineering analysis and design frameworks (Bisantz and Roth 2007) is that people in a system perform tasks with technology, so if tasks can be designed to match people and their skills in using technology, and if technology is designed to be usable and friendly, the task will turn out to be effective.

To understand this in the context of health care work, we posit that providers really perform tasks to create, share, remove, or otherwise manage information. Tasks, and the technologies designed to support them, provide a *means* for health care providers to get to information, their *end goal* (Figure 6.1).

While current design methods include information requirements gathering as part of their methodology (Nielsen 1994, 1997; Beyer and Holtzblatt 1998; Preece et al. 2002; Bonaceto and Burns 2007; Hoffman and Militello 2008), information is not the single and central unit of focus and analysis, as a task often is.

For example, current information requirements (Gorman 1995; Gorman and Helfand 1995; Shelstad and Clevenger 1996; Smith 1996; Gorman et al. 2000, 2002; Reddy and Dourish 2002; Reddy et al. 2002; Reddy and Spence

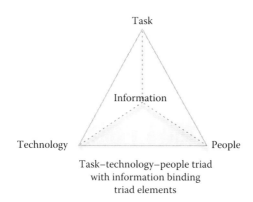

Task–technology–people triad
with information binding
triad elements

FIGURE 6.1
Information in the task technology triad—explains how information plays a role in the task technology triad.

2006, 2008; Pirolli 2007; Crowe et al. 2010) may identify the different fields (age, name, known allergies, severity level, insurance information, etc.) required in a patient information system interface. However, they do not always incorporate the intrinsic properties of the information entered in those fields and how those properties would afford the providers' use of that information.

Deep properties of information (Figure 6.2) such as the specific purpose of the information (e.g., making sure a discrepancy in medication order dosage is sorted out), information coordination and sharing needs (e.g., nurse sharing information on his/her progress with other providers), and information cues (e.g., highlighting information using colored markers) often do not constitute the task definition of the provider and the design specifications of the technology. For example, nurses creating medication error alert information may work around poor information support by duplicating the same information in many forms and places to share it with other providers—next to the patient's bed, verbal communication with providers, and on the whiteboard. The nurse may annotate the patient chart with text about the medication dosage using a red color; verbally communicate it to downstream providers; and handwrite the name of the medication and dosage on the whiteboard, circle it with a red marker, and attach an alert sticker for the patient. Note in this example the many low-level information creation activities that the nurse undertakes to further emphasize (to other providers) the importance and seriousness of the medication dosage change to minimize misses.

Although information plays a central role in the function of providers, there exists a mismatch between how providers create, use, and link information as the central tenet of their patient care work and how the design of information support system understands and represents their information needs.

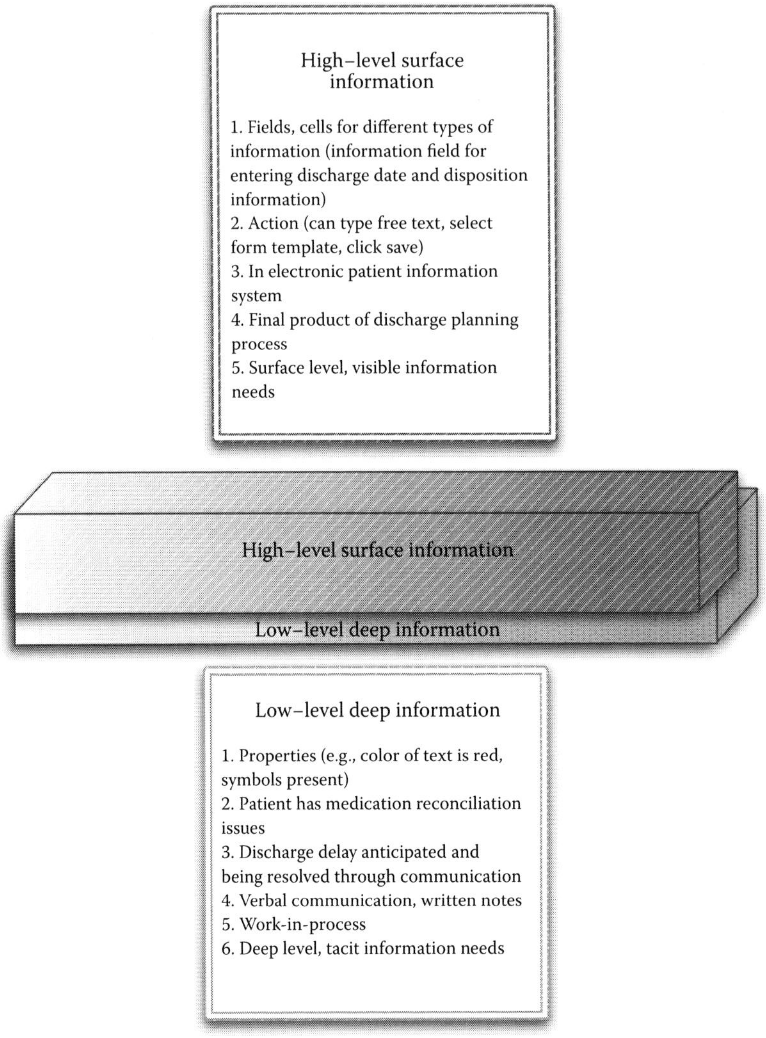

High–level surface
information

1. Fields, cells for different types of
information (information field for
entering discharge date and disposition
information)
2. Action (can type free text, select
form template, click save)
3. In electronic patient information
system
4. Final product of discharge planning
process
5. Surface level, visible information
needs

High–level surface information

Low–level deep information

Low–level deep information

1. Properties (e.g., color of text is red,
symbols present)
2. Patient has medication reconciliation
issues
3. Discharge delay anticipated and
being resolved through communication
4. Verbal communication, written notes
5. Work-in-process
6. Deep level, tacit information needs

FIGURE 6.2
High-level and low-level information.

The information trail model (Pennathur and Bisantz 2010; Pennathur 2011),
an information-based cognitive work analysis and design method and the
subject of this chapter, focuses its inquiry, analysis, and design centrally on
low-level details (Figure 6.2) regarding information creation, use, and transformation in a system—a design gap in the current methods—to develop
design recommendations for information support systems. Compared to
existing frameworks and methods in cognitive engineering and cognitive
work analysis, the primary difference and contribution of this method is

its central focus on information and related properties for research inquiry, analysis, and design. The target group who will benefit from this method includes human factors analysts, designers, and health care personnel. The intended benefits include (1) providing a new information systems framework for human factors analysts to approach a design or process issue; (2) providing user interface designers the insights and recommendations from understanding information as the central unit of work to help them develop innovative and user-responsive interfaces for health information systems; and (3) encouraging clinician involvement in the research and design process and ultimately benefiting the clinicians as the end user.

Information Trail Model for Cognitive Work Analysis and Design

The information trail model is an analysis and design method that has, as its central focus, the information people in the system create, use, and share. The model builds on the idea that people in a system self-organize using the information in their environment (Pennathur and Bisantz 2010; Pennathur 2011).

An information trail model essentially follows a "trail" of information creation, use, or sharing: people in a system create, use, and leave pieces or traces of information in the work environment. These information pieces, when joined together, produce an information trail (Pennathur 2011). The method emphasizes gathering deep, low-level information variables (see Figure 6.2 for a comparison between high- and low-level information) from the work environment. For further reading about the theoretical and conceptual framework guiding this method, please see the work of Pennathur (2011). In the rest of the chapter, we highlight the key features of the information trail model and provide tools for the practitioner for using the model.

Modeling Elements

The information trail model (Figure 6.3) consists of four main levels of information mapping: (1) state changes or unit operations (Taylor and Felten 1993) that link an individual's information piece with the trail of information, (2) individual information pieces, (3) a trail of information pieces linked together, and (4) influence of information on system dynamics (Meadows 1999). We briefly discuss the first three levels in the following paragraphs.

State Changes and Unit Operations

A state change occurs when an input to the system significantly changes into an output from the system, otherwise known as unit operation in the

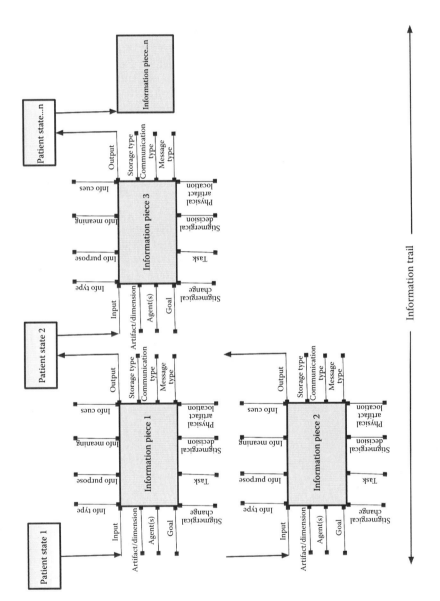

FIGURE 6.3
Overview of information trail model.

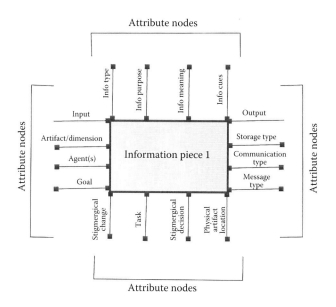

FIGURE 6.4
Information piece and attribute nodes.

sociotechnical systems framework (Figure 6.4; Taylor and Felten 1993). For example, when a new patient arrives in the ER, very little knowledge about the patient's condition exists. After the triage, a nurse asks specific questions about the patient's past medical history; new information about the patient emerges, initiating a state change in the extent of knowledge about the patient's condition.

Information Piece

An information piece refers to individual instances of information created or shared in a work system, e.g., different bits of information created when triaging a patient. An information piece has 16 attribute nodes (Figure 6.4) and can represent newly created information or information that already existed at the time of recording it. With each significant state change in a system (i.e., change from one significant unit operation to another), the corresponding information piece also transforms.

Each of the 16 nodes occurs in every information piece and represents important design characteristics (Table 6.1, Figure 6.4). These attribute nodes were selected based on their importance, relevance, and impact on the information under consideration using knowledge from literature. For more information about each of these attributes, please see the work of Pennathur (2011).

TABLE 6.1

Information Trail Attribute Nodes

	Description of Attribute Nodes
Input	Input to the system and every information piece. Each unit operation provides an input to the information piece.
Agent	Role responsible for creating, using, sharing, or transforming the information piece, e.g., nurse, physician.
Artifact/dimension	Tools and technologies used to create, use, share, or transform and represent it in a form in physical space.
Stigmergical change (marker vs. sematectonic [sema])	Active information-related changes and actions in the system due to special signs or cues left by the agents (marker). Passive information-related changes based on the current state of routine unit operations alone (sema).
Task	Tasks contained within the unit operations and for which the specific information piece was created, used, shared, or transformed.
Information type	Broad categorization of the information piece to understand the "type" of information supported, e.g., demographic information vs. discharge date. Information type is different from an information piece because there could be different information types captured within the same information piece, e.g., demographic, clinical information captured during triage.
Information purpose	The reason for creating, using, sharing, or transforming the specific information piece.
Information meaning	Meaning of the information created, used, shared, or transformed under a specific information piece. Meaning can either be directly understood or may be tacit only to the person who created it or local to the system.
Communication	Type of communication occurring between agents in the system for this specific information piece. Communication can directly and intentionally occur between agents, or one agent could create information that is only meant for them, but could unintentionally be useful for other agents in the system.
Information storage type	Capturing whether the information piece needs to be stored permanently or temporarily and if the tool or technology representing it supports the storage type.
Message type	Intent for action of the agent in creating the information piece, e.g., asking a question, requesting information
Information cues/symbol/ color	Annotations that people add to information to make it more salient, important, and usable for their own purpose.
Individual goal	Reason why the agent created, used, or shared the specific information piece. This may be the same or different as the system level goal for the information piece.
Physical location of artifact	The location of the tool or technology where the information piece is represented.

(continued)

TABLE 6.1 (Continued)

Information Trail Attribute Nodes

	Description of Attribute Nodes
Stigmergical decision (quantitative vs. qualitative)	The nature of the object, entity, or characteristic determining whether decisions are made quantitatively or qualitatively.
Output	The resulting outcome from the transformation of the information piece, which feeds into the next unit operation, e.g., if a "not seen patient" is the input, then the "triaged" patient may be the output of a "triage information piece."

Information Trail

Once assembled in their order of occurrence, the individual information pieces are integrated to produce an information trail. An information trail re-represents the different ways people perform work using information as the basis to achieve their goals (Figure 6.4).

In this chapter, examples of patient transition across the OR, ICU, and discharge in the inpatient unit illustrate use of the information trail model for health care systems.

Information Trail Modeling Phases

Three distinct phases constitute the information trail model (Figure 6.5). The first phase identifies the specific health care subsystem of interest. The processes, technologies, and people involved in a specific health care subsystem are examined. The next phase involves modeling the information created, shared, or managed in each of the processes identified in phase 1. The third phase includes quantitative and qualitative analyses of data. In this chapter, we discuss only phase 2 in detail. A brief illustration of phase 2 is provided in Figure 6.6 because phase 1 closely follows the sociotechnical modeling definition of unit operations and state changes (Taylor and Felten 1993), and phase 3 includes qualitative and quantitative analyses that are briefly discussed.

Phase 2: Information Modeling

In the information-modeling phase (Figure 6.7), we gather data on low-level information and its associated properties represented in each of the 16 attribute nodes of the information trail model (see Figure 6.5 for a brief review of these 16 nodes). Observations of the system and shadowing of providers help model information creation, use, sharing, and transformation.

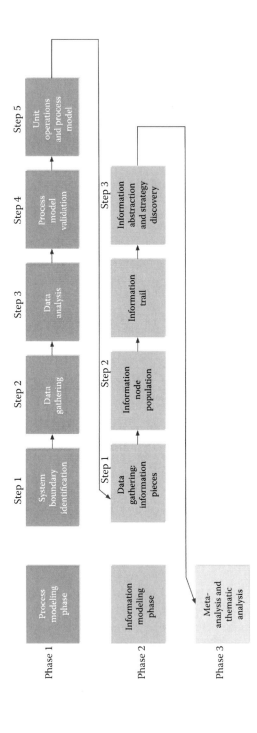

FIGURE 6.5
Overview of phases in information trail model.

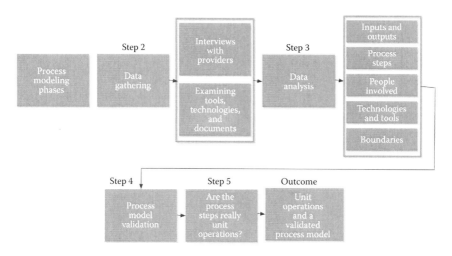

FIGURE 6.6
Data-gathering steps leading to analysis and unit operations identification.

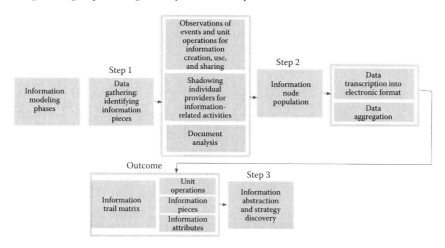

FIGURE 6.7
Overview of information modeling phase.

The steps in information modeling include

1. Synthesis and aggregation of the individual instances and pieces of information identified along with their attributes

2. Linking the individual instances of information to form an information trail

3. Tagging the attribute nodes with more generic abstract properties of information (e.g., tagging red color in information cues attribute node with salience as an abstracted property)

Step 1: Data Gathering: Information Piece Identification at Significant States

The first step identifies individual instances of low-level information created, used, shared, or managed at the identified unit operations (Figure 6.7). Individual instances of information can range from creating a sheet with handoff communication information (i.e., multiple details on the same sheet but for the same purpose) to updating the progress note with the day's tests (i.e., a single detail about the tests).

As we identify the instances of information based on the unit operations already identified, we do not focus on the amount of information but rather

TABLE 6.2

Observation Checklist

	Corresponding Information Trail Nodes or Properties
Where is the information coming from? Is it newly created, or did someone leave an indication or trace of the information that is now being used or created?	Input, stigmergical change, information cues, task
What tools are being used to create or share the information?	Artifact/dimension, physical location
Who is creating and using the information and whom do they intend it for?	Agent, communication, information meaning, message type
Why is this information being created and how is it likely to be used?	Information purpose, goal
What is the task in which this information is applicable?	Task
Is the information in numerical form with scalar and quantitative information or is it qualitative assessment of a situation? (e.g., 120/80 blood pressure vs. skin was pale, eyes were drooping, was visibly in pain)	Stigmergical decision
Are there any information cues, special highlights that make the information salient?	Information cues
Is the meaning of the information directly comprehensible, or is it implicit to the local culture?	Information meaning
Is the information stored temporarily or permanently?	Storage type
How is the information communicated? Is the communication/sharing intentional?	Communication
How has this information piece changed the input? What has changed, and where is the information going?	Output
Are any other information pieces being created or used simultaneously in the system for the same or different unit operation?	Helps identify information that is created or used simultaneously
Are any other information pieces being created or used repeatedly in the system for the same or different unit operation?	Helps identify information that is created or used repeatedly
Is this a work-around? Is this a work practice or a tool that is user-developed? Why?	Capturing design affordances and user strategies

on the purpose of information binding those instances together at a particular time point. The outcome resulting from this data-gathering phase includes a collection of low-level information instances created, used, or shared in the system at each of the unit operations, and links between them showing a relationship to the global goal of the system. Additionally, qualitative categories of strategies that providers use to manage information also result from this phase.

Observations and Shadowing

Observations and shadowing can be conducted for each unit operation identified to capture low-level information created, used, shared, or managed by providers. Whereas observations can capture data about low-level information creation, use, sharing, and management from more focused events (e.g., verbal handoff report), shadowing will help understand details of provider activities more thoroughly during routine work (e.g., updating progress note). The information characteristics that are essential to capture during observations are provided in Table 6.2.

When observing information-related activities, capturing the actual information created assumes less importance than capturing the type of information and properties associated with this information. For example, it is unnecessary, and sometimes even impossible, to capture patient-related information such as "78-year-old Henry." Instead, the type of information, which in this case describes age (in numeric), name, and such demographic information, suffices our needs.

Additionally, the order of information transformation activities may prove important. Information can repeatedly get created in a system or created just once. For example, a fellow creates and maintains a progress note for a patient everyday. However, a provider generates a discharge summary only once before the patient leaves the hospital.

Capturing information that is created or used at the same time point through representative snapshots that illustrate concurrent information creation or use can also provide value.

Document Analysis

Document analysis for information modeling can provide insights on the type of artifacts used for information transformation activities. Additionally, analyzing documents helps identify current structure, content, purpose, and a variety of tools used to support information transformation.

Step 2: Information Node Population

Synthesis and Aggregation: Matrix Mapping

Data from observations, shadowing, and document analysis undergo transcription as necessary (Figure 6.8). The following steps are involved in

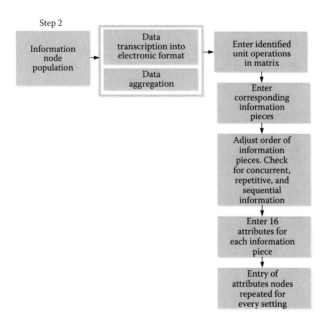

FIGURE 6.8
Steps in data aggregation and matrix mapping.

aggregating the data collected on low-level information creation, use, and sharing activities.

1. The first step in combining and aggregating data involves data organization. All significant unit operations identified in phase 1 are entered into an electronic matrix.

2. Under each of the unit operations, we enter their corresponding information pieces identified in phase 2. For these information pieces, the cells corresponding to the 16 attribute nodes are left blank initially (see Figure 6.9). Categorizing by roles can make the analysis easier. For example, categorizing the matrix by unit operations and information pieces of the nurse, physician, fellow, resident, and so on can help the initial data organization.

3. Next, we adjust the matrix to capture the order in which the information pieces occurred. That is, concurrent or repetitive information and sequences, if any, need to be corrected in the electronic matrix. Depending upon the number of similar settings studied, the level of integration will differ. For example, if the study includes only one ICU or inpatient ward, we require only data entry and a moderate level of synthesis in contrast to a study comparing two floor units. Multiple settings (e.g., ICU1 followed by ICU2) necessitate repetition of step 1. The iterative nature of data entry helps aggregation of

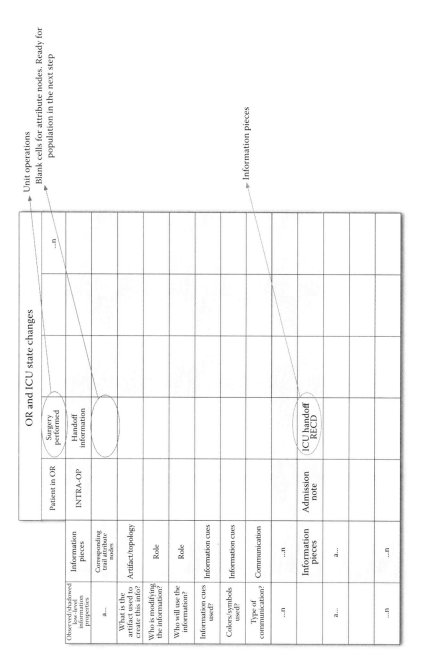

FIGURE 6.9
Unit operations and information pieces entry.

Information property based on observation/shadowing from one setting

OR and ICU state changes				
Observed/shadowed low-level information properties	Information pieces	Patient in OR	Surgery performed	...n
a...	Corresponding trail nodes	INTRA-OP	Handoff information	
What is the artifact used to create this info?	Artifact/topology	Paper	Paper	
Who is modifying the information?	Role	Anes. resident	Anes. fellow	
Who will use the information?	Role	Physician, nurse	Physician, nurse	
Information cues used?	Information cues	Highlighted	Circled	
Colors/symbols used?	Information cues	Red	Blue	
Type of communication?	Communication	Direct	Indirect	
...n	...n			
	Information pieces	Admission note	ICU handoff RECD	
a...	a...			
...n	...n			

FIGURE 6.10

Information attribute node entry.

information pieces across settings and allows for the addition of new information pieces as they occur.

4. Next, for each information piece, the qualitative data from observations, shadowing, and examination of documents on low-level information properties are entered into the blank cells of the 16 attribute nodes (Figure 6.10). For example, in OR1, during the unit operation "surgery performed," for the information piece "handoff information," and for the attribute node "artifact used," the data could indicate that "paper" was used and is entered into the blank cell for the "artifact used."

5. Data for multiple settings, if any, need aggregation. For example, in OR2, the data could indicate that "electronic medical record" was used for the same "surgery performed"–handoff information"–"artifact used node." To aggregate, we combine the "electronic medical record" with "paper form" for the attribute node "artifact used" (see Figure 6.11 for an example). Similar data across participants and settings are retained, and new responses are carefully combined, so that aggregation of unlike data does not occur. The resulting information trail matrix represents the unit operations, information pieces, and their associated properties.

In summary, it helps to visualize the information trail as a route map providing all the information-related activities that providers ever did for the patient before, during, and after their hospital stay. The route map when expanded also gives specific directions or details about the properties of the low-level information. An information trail matrix provides both a bird's eye view of information generated and used, and a detailed view of the low-level information in the system.

Step 3: Information Abstraction and Strategy Discovery

This step involves abstracting the attribute nodes in the information trail matrix (see Figure 6.12 for an example). We examine each response in the unit operation–information piece–information trail node combination to develop categorical indicators. For example, for the attribute node "artifact used," if the responses were "I use electronic mostly for putting in orders," "I use paper notes just during rounds," and "I use paper notes for my workflow," categories such as "workflow/manual" and "documentation/electronic" can be developed using thematic content analysis.

Abstracted information trail nodes provide knowledge about the strategies that providers use to deal with information and may include accessing, creating, sharing, processing, transferring, and removing information. Abstraction also helps to develop context-free strategies, which helps to compare strategies for information use across different domains.

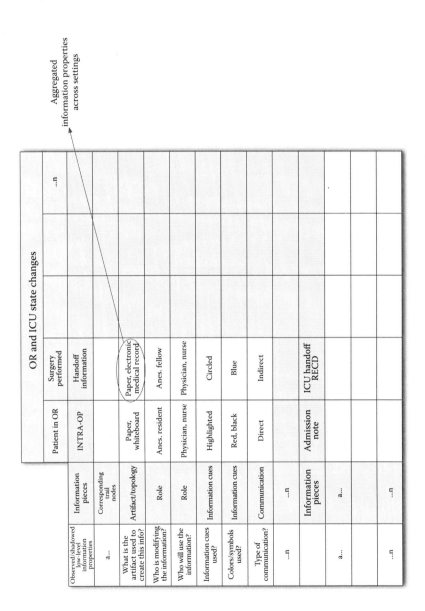

FIGURE 6.11
Information attribute node aggregation across settings: resulting information trail matrix.

Interview questions	Information pieces	Patient in OR	Surgery performed			...n
	Corresponding trail nodes	INTRA-OP	Handoff information		Paper	
a...						
What is the artifact used to create this info?	Artifact/topology	Manual artifact	Manual primary electronic updater			
Who is modifying the information?	Role	Anes. resident	Anes. fellow			
Who will use the information?	Role	Physician, nurse	Physician, nurse			
Information cues used?	Information cues	Salience enhancing	Workflow indicator			
Colors/symbols used?	Information cues	Severity indicator	Shift change			
Type of communication?	Communication	Direct	Indirect			
...n	...n					
	Information pieces	Admission note	ICU handoff RECD			
a...	a...					
...n	...n					

OR and ICU state changes

Abstracted information properties across settings

FIGURE 6.12
Information trail abstraction.

Transition from OR to ICU: Case Example of an Information Trail Model

This section discusses a representative example of a handoff from OR to ICU. We explain each attribute node in the information trail model through information pieces derived from either anecdotal observations or reports and some hypothetical scenarios. Note that each information piece discussed in the following example would have 16 attribute nodes, but for reasons of space, we discussed only four information trail attribute nodes per information piece (Figure 6.13).

In a typical analysis, the patient serves as the input, and information transformation modeling continues until the patient exits the system as the output. In our example, we will only explore different nodes of the information trail when Henry, a 63-year-old patient undergoing cardiac surgery in the cardiac OR, transfers to the ICU.

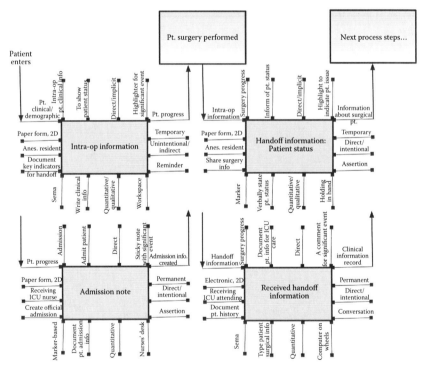

ICU handoff: Information trail model illustration

The final outcome from qualitative analyses will be similar to this example representation but with generalized representative information properties in condensed trails.

FIGURE 6.13
Transition from OR to ICU information trail model.

These four instances depict four key representative snapshots of information creation and use events for the handoff process. These information transformation activities will continue until and after Henry leaves the hospital system. However, description of only these four instances follows.

Intra-Op Clinical Information

Input   The patient and the corresponding clinical and demographic information serve as the input to the intra-op information piece as well as to the entire information trail. Henry's information serves as the trigger for all other information creation and use activities in the trail.

Agent   The anesthesiology resident monitors Henry's vitals and records relevant patient information during the surgery. The agent node captures the role of the person creating or using the information. An agent in the information trail can denote a human or sometimes automation, depending on the context.

Artifact/dimension   The anesthesiology resident records clinical and demographic information about Henry on a paper form. The artifact/dimension node models the artifacts used for information creation or use. The artifacts used may vary from one information piece to the next. The information written on the paper form is a two-dimensional text. The dimension of the information/artifacts suggests the different possibilities for interaction.

Goal   The goal node models the individual goal of the agent. The goals will change from one agent to another as well as from one state to another. In the intra-op information piece, the goal of the anesthesiology resident involves documenting key indicators for Henry both for handing off information and for their individual memory jogging about Henry. Note that this information recording does not end with a one-time point event and will continue across time. The anesthesiology resident enters the information as the surgery progresses. After the surgery ends, Henry is transferred to the ICU.

Admission Note

As the surgery nears completion, the circulating nurse contacts the ICU, and the receiving team in the ICU starts preparing for the arrival of Henry. An admission note is created in a paper form with Henry's basic clinical and demographic information.

Stigmergical Change   Actions in a system can stem from the current state of the system (sematectonic stigmergy) or from special markers in the information left by people (marker-based; Bonabeau et al. 1999; Parunak and Brueckner 2003; Parunak 2005). This node models how the providers in the system create or use dynamic information. For example, information that the OR anesthesiology resident enters in the clinical information system helps generate the admission note. A provider transfers this information to the admission note form. Therefore, this example of a marker-based stigmergy illustrates how information traces left in the system help create new information.

Task   The task node represents the task underlying the information created or used. In this case, the task of documenting Henry's information in the admission note introduces Henry into the ICU system workflow.

Stigmergical Decision   Information created by humans often helps in making decisions either based on the quantitative aspects or the qualitative features (Bonabeau et al. 1999; Parunak and Brueckner 2003; Parunak 2005). While most real-world decisions can use both quantitative and qualitative aspects, this classification helps us understand the specific nature of information that leads to certain decisions. In this case, the ICU nurse knows the information that he or she has to record based on the structured items present in the admission note form (quantitative).

Physical Artifact Location and Proximity to Agent Space   The physical location of the artifact determines the possibilities for action (multiple people viewing the same information), control of information (access issues), and constraints in information creation. In this case, the admission note is located by the ICU nurse's desk for easy reference and information documentation.

Hand-Off Information
As the provider prepares Henry for the ICU stay, the anesthesiology resident, the OR handoff team, and the ICU receiving team assemble in Henry's room. The anesthesiology resident shares information about Henry with the rest of the team.

Information Type   The type of information models the categories of information created/used in the system. For example, in this case, the anesthesiology resident shares clinical and demographic information with the receiving team.

Information Purpose   The purpose of information in the admission note is to officially record Henry's entry to the ICU. Note that although the information purpose node may seem to overlap with the goal node, the information purpose node captures the purpose of information, whereas the goal node captures the goal of agents.

Information Meaning   Sometimes, everyone can easily interpret the meaning of information, or it can remain implicit only to the person who created it or those who share the same work. The admission note information offers direct interpretation by all of the other providers both in the ICU and the OR.

Information Cues People often qualify information to increase the salience, make information easily understandable, or filter and make the information more usable. The ICU nurse notes that the patient has a preexisting condition that may need a slightly different treatment plan and something that has the potential to cause safety issues. Therefore, the ICU nurse wants to bring attention to this fact and highlights that particular information with a highlighter, so that the other providers do not miss this information. Note how this simple cue can make the tasks of other providers easier and helps prevent major safety issues for the patient.

Hand-Off Information Received

As the anesthesiology resident shares information, the ICU resident records information in a clinical information system. The resident does not enter all the information shared by the anesthesiology resident. The ICU resident enters only information pertinent to the ICU care and asks specific questions that may support the same. When the ICU team receives the information, at least two separate important information pieces can exist in reality. One involves the actual listening/questioning and receiving of the information, and the second involves the record of the information on an artifact. In addition, the ICU resident records information as the anesthesiology resident shares the information. Therefore, these separate information pieces occur almost concurrently.

Message Type Every time information creation occurs, a message is implied. A message can include a request for something, a question, a confirmation/assertion, or a conversation. The ICU resident engages in a conversation with the OR team and transfers the information to their ICU system.

Communication Type Communication of information from the sender to the receiver occurs either intentionally with a purpose or unintentionally, which benefits the receiver. Similarly, the sender directly or indirectly communicates information to the receiver through artifacts. The ICU team upon receiving handoff communication intentionally records and communicates in the clinical information system to make other providers aware.

Storage Needs/Storage Affordance The information created need not always require permanent storage in the system. People often create and store information temporarily to serve a purpose and remove it from the system. The artifact where the information is stored also needs to afford such storage. For example, a small sheet of blank paper used to note down information may afford temporary storage and removal from the system easily, whereas a huge binder with other patient information cannot easily store or remove temporary information.

Output The ICU resident's information entry results in an output of information record of the patient in the ICU clinical system. An output in every information piece results in a state change and serves as the input for the next information piece. Several intermediate outputs and concurrent outputs trigger a state change. The final output of the information trail satisfies the overall system goal of the patient being transferred from the ICU.

Phase 3: Meta-Analysis, Thematic Analysis, and Prioritization of Design Needs

Meta-analysis can determine the extent of occurrence of an information node. Only the nodes with a dichotomous response category are included in the meta-analysis.

For example, if we had identified a total of 100 unique information pieces in a system, the meta-analysis may reveal that out of those 100 information pieces, 75 had temporary storage and 25 had permanent, indicating that 75% of the information underwent constant transformation or removal, and only 25% required permanent storage.

On the other information nodes such as agent (e.g., nurse, physician), artifact/dimension (e.g., electronic, manual, none), message type (e.g., confirmation, question, request), and physical artifact location (e.g., near nurses' station, near whiteboard), count analysis of similarly occurring items will reveal trends. For example, if meta-analysis reveals that 65 of 100 information pieces had some type of manual artifact (i.e., paper, whiteboard, notecard), 30 were electronic, and 5 neither, we can identify if the temporary storage of information relates to the usage of manual artifacts.

While meta-analysis provides a quantitative indicator of information and its properties in the health care system, a qualitative analysis can identify workflow themes considered significant by providers to help generate design priorities.

Discussion

Information modeling has contributions to theory, methods, and design (Pennathur 2010). Some of the significant contributions are discussed.

Concurrency of Information in the System

Information in large complex systems is not created in isolation. Information creation can occur concurrently (Pennathur 2010) and either for joint use or for parallel processing of information. Pieces of information created or used in the system are interdependent with each other. For example, as the nurse in the OR calls the ICU about the anticipated end time of the patient's surgery, the ICU nurse may create the admission note.

The information trail model captures the concurrency in information creation, use, sharing, and transformation in the system. Concurrency helps denote information that may need joint use, integration, and the temporal aspects of information use for design. A design feature in this instance could be an automatic extraction of patient information from the OR system at the time of the phone call to help the ICU nurse create a comprehensive admission note.

Repetition of Information

Information sometimes needs repeat creation or use. Although the same information may not get created or used repeatedly, similar types of information may repeat in a system (Pennathur 2010). Capturing repetition in information for design can help reduce workload, errors, and unintended consequences.

If a piece of information needs to repeat, then the features in the design need to effectively support repetition without introducing additional redundancy of information. For example, having a copy/paste button for repetitive text will reduce effort in recreating the same information but instead can lead to redundant and temporally and situationally inappropriate information for the patient and make the information error-laden (see Chapter 9 of this book for a related discussion on information repetition in electronic health systems).

Information Visualization

Information is abundant in the health care system, but its utility for decision making and patient care delivery remains limited. While our capability for storing and processing health care information has improved over the years, making sense of this vast information database can facilitate the health providers' understanding of patient conditions and will improve communication in the entire system. Information trail modeling documents the work practices of health care personnel who demonstrate the need for effective visualization of information through their everyday creation and use of visual cues and annotation in information and artifacts. Such visual cues and annotations can be combined with information needs derived from other cognitive engineering models (such as the work domain models as demonstrated in Chapter 4 of this book) in creating information displays.

Storage of Information

Information is transient and may not always require permanent storage. Temporary storage of information may be warranted in situations requiring confidential information, information that is used only on a day-to-day basis, information that is specific to workflow, etc. The choice to temporarily store information is also dependent on whether the artifact affords such storage, and hence becomes an important design implication to consider. Leaving out those options may prompt the user to develop work-arounds that support temporary and secure storage.

Strategies in Information Transformation

Strategies that providers use to deal with information, regardless of the technology, tools, or official practices already in place in the system, help us understand what providers really need from information systems and their actual workflow. While the strategies may be local to a system (Gordon 2004; Pennathur 2010), with customizable systems, these are no longer infeasible to implement in design. Generic strategies for information use or creation, in the abstracted form when used by providers to deal with information, can be captured in the design, so the final design incorporates the why and need not necessarily replicate the form of how the provider carried out the strategy. A unique contribution from our information modeling approach includes derivation of strategies.

Markers and Signs

Employing markers and signs in the creation, use, and sharing of information indicates the need to actively or passively engage in acts of cooperation and coordination, known as stigmergy. Stigmergy, a special case of self-organization, is where people leave cues in the environment prompting further actions, sometimes even without direct communication (Bonabeau et al. 1999; Susi and Ziemke 2001; D'Angelo and Pagello 2002; Parunak and Brueckner 2003; Shell and Mataric 2003; Parunak 2005). In workplaces where teams are the norm, such as in health care, identifying such needs for cooperation and coordination via information provides insights on designing sharing, communication, and coordination support within a technology. Chapter 3 of this book discusses related aspects of team support.

Conclusions

The information modeling approach in health care systems allows understanding information creation, use, sharing, and transformation and how information support would be essential in any new design or redesign of systems. Information modeling approach addresses the fundamental question of how people in the system perform their work with information as the binding element among technologies, tasks, and people. Information trail modeling will result in a richer understanding of the current system, its work practices, and information needs, paving the way for a more robust usable health care information system.

Acknowledgment

Pennathur is supported by a mentored grant from the National Library of Medicine, NIH (1K99LM011309-01).

References

Arora, V. (2005). Communication failures in patient sign-out and suggestions for improvement: A critical incident analysis. *Quality and Safety in Health Care, 14*(6), 401–407. doi: 10.1136/qshc.2005.015107.

Ash, J.S., Berg, M., and Coiera, E. (2004). Some unintended consequences of information technology in health care: The nature of patient care information technology system-related errors. *Journal of the American Medical Informatics Association, 11*(2), 104–112.

Barker, K.N., Flynn, E.A., Pepper, G.A., Bates, D.W., and Mikeal, R.L. (2002). Medication errors observed in 36 health care facilities. *Archives of Internal Medicine, 162,* 1897–1903.

Beyer, H., and Holtzblatt, K. (1998). *Contextual Design: Defining Customer-Centered Systems.* San Diego, CA: Academic Press.

Bisantz, A., and Roth, E. (2007). Analysis of cognitive work. *Reviews of Human Factors and Ergonomics, 3*(1), 1–43.

Bonabeau, E., Theraulaz, G., and Deneubourg, J.-L. (1999). *Swarm Intelligence: From Natural to Artificial Systems.* New York: Oxford University Press.

Bonaceto, C., and Burns, K. (2007). A survey of the methods and uses of cognitive engineering. In R.R. Hoffman (Ed.), *Expertise Out of Context: Proceedings of the Sixth International Conference on Naturalistic Decision Making.* New York: Lawrence Erlbaum Associates.

Brennan, T.A., Leape, L.L., Laird, N.M. et al. (2004). Incidence of adverse events and negligence in hospitalized patients: Results of the Harvard Medical Practice Study I. *Quality and Safety in Health Care 13*(2), 145–152.

Calleja, P., Aitken, L.M., and Cooke, M.L. (2011). Information transfer for multi-trauma patients on discharge from the emergency department: Mixed method narrative review. *Journal of Advanced Nursing, 67*(1), 4–18.

Croskerry, P. (2003). The importance of cognitive errors in diagnosis and strategies to minimize them. *Academic Medicine, 78*(8), 775.

Crowe, S., Tully, M.P., and Cantrill, J.A. (2010). Information in general medical practices: The information processing model. *Family Practice, 27*(2), 230–236. doi: 10.1093/fampra/cmp102.

D'Angelo, A., and Pagello, E. (2002). Using stigmergy to make emerging collective behaviors. Paper presented at the *Proceedings of the Eighth Conference of the Italian Association for Artificial Intelligence,* Siena, Italy.

Gordon, A.S. (2004). *Strategy Representation: An Analysis of Planning Knowledge.* Mahwah, NJ: Lawrence Erlbaum.

Gorman, P. (1995). Information needs of physicians. *Journal of the American Society for Information Science, 46*(10), 729–736.

Gorman, P., Ash, J., Lavelle, M. et al. (2000). Bundles in the wild: Managing information to solve problems and maintain situation awareness. *Library Trends, 49*(2), 266–289.

Gorman, P.N., and Helfand, M. (1995). Information seeking in primary care. *Medical Decision Making, 15*(2), 113.

Gorman, P., Lavelle, M., Delcambre, L., and Maier, D. (2002). Following experts at work in their own information spaces: Using observational methods to develop tools for the digital library. *Journal of the American Society for Information Science and Technology, 53*(14), 1245–1250.

Halbesleben, J., Jonathon, R.B., Savage, G.T., Wakefield, D.S., and Wakefield, B. (2010). Rework and workarounds in nurse medication administration process: Implications for work processes and patient safety. *Health Care Management Review, 35*, 124. doi: 10.1097/HMR.0b013e3181d116c2.

Hoffman, R.R., and Militello, L.G. (2008). *Perspectives on Cognitive Task Analysis*. New York: Psychology Press.

Kohn, L.T., Corrigan, J.M., and Donaldson, M.S. (Eds.) (1999). *To Err is Human: Building a Safer Health System*. Washington, DC: National Academies Press.

Koppel, R., Wetterneck, T., Telles, J.L., and Karsh, B.-T. (2008). Workarounds to barcode medication administration systems: Their occurrences, causes, and threats to patient safety. *Journal of the American Medical Informatics Association, 15*(4), 408–423.

Leape, L., Bates, D., Cullen, D. et al. (1995). Systems analysis of adverse drug events. *Journal of American Medical Informatics, 274*(1), 35–43.

Leape, L.L., and Berwick, D.M. (2005). Five years after to err is human. *JAMA: The Journal of the American Medical Association, 293*(19), 2384.

Meadows, D. (1999). *Leverage Points: Places to Intervene in a System*. VT: Sustainability Institute.

Nielsen, J. (1994). Ten usability heuristics. Available at http://www.useit.com/papers/heuristic.

Nielsen, J. (1997). Usability testing. In G. Salvendy (Ed.), *Handbook of Human Factors and Ergonomics* (2nd ed.). New York: John Wiley & Sons, Inc.

Parunak, H.V.D. (2005). A survey of environments and mechanisms for human-human stimergy. *Proceedings of the 2nd International Conference on Environments for Multi-Agent Systems*, 163–186.

Parunak, V.D., and Brueckner, S. (2003). Tutorial on engineering self-organizing applications. Paper presented at the *AAMAS Tutorial on Engineering Self-Organizing Applications*.

Pennathur, P.R. (2010). Information transformation and artifact use in cognitive work systems: Implications for technology transition and design. Unpublished PhD dissertation, University at Buffalo, State University of New York, Buffalo, NY.

Pennathur, P.R. (2011). An information trail model for capturing human behavior in artifact creation and use in complex work systems. *Theoretical Issues in Ergonomics Science, 14*(4), 311–359.

Pennathur, P.R., and Bisantz, A.M. (2010). A novel information trail model for information transformation in cognitive work systems. *Proceedings of the Human Factors and Ergonomics Society 54th Annual Meeting*, San Francisco.

Pirolli, P. (2007). *Information Foraging Theory: Adaptive Interaction with Information.* New York: Oxford University Press.

Preece, J., Rogers, Y., and Sharp, H. (2002). *Interaction Design: Beyond Human-Computer Interaction.* New York: John Wiley & Sons, Inc.

Reddy, M., and Dourish, P. (2002). A finger on the pulse: Temporal rhythms and information seeking in medical work. *Proceedings of the 2002 ACM Conference on Computer Supported Cooperative Work,* 344–353.

Reddy, M., Pratt, W., Dourish, P., and Shabot, M. (2002). Asking questions: Information needs in a surgical intensive care unit. Paper presented at the *Americal Medical Informatics Association Annual Symposium,* Washington, DC.

Reddy, M., and Spence, P.R. (2006). Finding answers: Information needs of a multidisciplinary patient care team in an emergency department. *AMIA Annual Symposium Proceedings, 2006,* 649.

Reddy, M.C., and Spence, P.R. (2008). Collaborative information seeking: A field study of a multidisciplinary patient care team. *Information Processing & Management,* 44(1), 242–255.

Shell, D.A., and Mataric, M.J. (2003). On the use of the term "stigmergy." Paper presented at the *Proceedings of the Second International Workshop on the Mathematics and Algorithms of Social Insects,* Atlanta, Georgia.

Shelstad, K.R., and Clevenger, F.W. (1996). Information retrieval patterns and needs among practicing general surgeons: A statewide experience. *Bulletin of the Medical Library Association, 84*(4), 490.

Shulman, R., Singer, M., Goldstone, J., and Bellingan, G. (2005). Medication errors: A prospective cohort study of hand-written and computerised physician order entry in the intensive care unit. *Critical Care, 9*(5), R516–R521. doi: 10.1186 /cc3793.

Smith, R. (1996). What clinical information do doctors need? *BMJ: British Medical Journal, 313*(7064), 1062.

Susi, T., and Ziemke, T. (2001). Social cognition, artefacts and stigmergy: A comparative analysis of theoretical frameworks for the understanding of artefact-mediated collaborative activity. *Cognitive Systems Research, 2*(4), 273–290.

Taylor, J.C., and Felten, D.F. (1993). *Performance By Design: Sociotechnical Systems in North America.* New Jersey: Prentice Hall.

Williams, R., Silverman, R., Schwind, C. et al. (2007). Surgeon information transfer and communication: Factors affecting quality and efficiency of inpatient care. *Annals of Surgery, 245*(2), 159. doi: 10.1097/01.sla.0000242709.28760.56.

Wilson, R., Harrison, B., Gibberd, R., and Hamilton, J. (1999). An analysis of the causes of adverse events for the quality in Australian Health Care Study. *The Medical Journal of Australia, 170,* 411–415.

Wilson, R., Runciman, W., Gibberd, R. et al. (1995). The quality in Australia health care study. *The Medical Journal of Australia, 163,* 458–471.

7

Support for ICU Clinician Cognitive Work through CSE

Christopher Nemeth, Shilo Anders, Jeffrey Brown,
Anna Grome, Beth Crandall, and Jeremy C. Pamplin

CONTENTS

Introduction

Cognitive systems engineering (CSE) has been proven to be useful in reveal-ing key aspects of operator behavior as they pursue goals in complex work domains, providing the foundation for the development of solutions that are ecologically valid. Health care work settings, particularly the intensive care unit, present one of the most challenging work domains for a researcher to study. Cognitive engineering methods (Hollnagel and Woods 1983; Woods and Roth 1988; Roth et al. 2002; Militello et al. 2010) can be applied to under-stand characteristics of complex work domains such as the ICU as well as the behavior of workers including clinicians and their support staff. The use of CSE methods makes it possible to identify key traits of health care work set-tings, such as decisions clinicians make, obstacles clinicians face, and initia-tives they take to overcome these obstacles in their efforts to restore patients to the best possible health. CSE methods also have the potential to enable workers to better understand their unit's performance and more successfully adapt to unforeseen challenges—in other words, to be *resilient*.

This chapter describes a project using CSE methods that is underway at a burn intensive care unit (BICU) in a major military medical center. This project will develop an ecologically valid computer-based cognitive artifact (Hutchins 2002) that will support individual and clinical team decisions and communication.

Background

The study of health care relies on the use of proven methods by qualified researchers. This is because work at the sharp (operator) end of health care is (among other traits) dense, time-pressured, and complex. Expert workers can find it difficult to be objective observers of their own activities and work settings. Because of this, studying one's own system may yield conclusions that are logical but may also miss deeper issues. Attention in such studies often focuses on a single theme while excluding the many elements that interact with each other to produce a collective result—its context.

For example, *closed claims reviews* that conclude that error elimination will remove "error causes" ignore the complex pressured context that molded each event. It assumes that a claim will contain all of the information that needs to be known about an adverse outcome. It also presumes to know what caused that outcome, that it was caused by an "error," and that its cause can be "eliminated."

Retrospective records review relies on historical documentation in order to draw conclusions about care and its related risks. But records hold little of the context, speculation, deliberation, and complex trade-off decisions that typically mold any significant event.

Voluntary reporting systems have been touted as tools to incorporate error reporting and analysis into the culture of medicine (Plews-Organ et al. 2004). However, voluntary reporting fails to note how the approach is vulnerable to social and organizational influences.

Clinical discussions of patient safety often review how effective a single diagnostic or therapeutic intervention is without taking other factors into account that would affect outcomes in actual practice. For example, Shojania et al. (2001) tested the use of a single item to prevent infections: a maximum sterile barrier when placing intravenous catheters. Some clinicians attempt to make system analysis easier by bounding the problem through selection and management of a single variable. Kyriacou et al. (1999), for example, sought to measure and reduce the length of stay in the emergency department. Some clinicians have applied methods such as workload assessment to the ED, but they found that the level of effort that is required makes it difficult to routinely use it as a measurement tool (Levin et al. 2006). Others have imported measures from other sectors to measure a single aspect of ED operation. For example, France and Levin (2006) used the notion of "system complexity" to determine safe capacity during care demand surges but conceded that phenomena such as interruptions need to be added.

Research that does not adequately detect or understand these issues diverts valuable resources into low-yield efforts. Research that reveals context will grasp the constraints that shape opportunities and risks in practice, curb the influence of hindsight and outcome bias, and yield valid solutions that gain traction in actual work settings (Wears and Nemeth 2007). A current intensive care unit study provides an illustration of how the use of CSE makes that possible.

Research Design and Methods

Our research team is completing the first part of a three-year study to develop a computer-based cognitive aid that supports cognitive work and communication. While it is still in its early stages, it can serve as an example of CSE's value in health care. We discuss the CSE approach in this chapter in the context of our work on a prior project that described quality standards for how to conduct CSE research.

Quality

Nemeth et al. (2011) described the use of CSE in a Navy-funded project that demonstrated how to use the CSE approach in the context of the Department of Defense acquisition process. The project's results would be used by government staff members and contractors who have no prior CSE training or experience. The approach needs to be used well to produce useful results. How would the new users know what that is? The team conceived of "reasonable scientific criteria" as a way to guide new users through CSE in a manner that is scientifically rigorous and that links design recommendations directly to operator needs. Using steps in the CSE process, the team considered the goals and activities at each stage, case studies from the literature that exemplified each stage, and ways that performance and scientific rigor could be evaluated at each stage. In order to do that, the team considered three questions:

1. What reliability/validity criteria are important and reasonable to apply to CTA data?
2. What are the standards of practice, and what needs to be done to meet those standards?
3. How can a rigorous process be created and followed while also being open to discovery with respect to process and outcome?

Answers to these questions identified a set of quality standards for each stage of the CSE process (Table 7.1) from Nemeth et al. (2011) that can also be applied to research in the health care context.

In the section "Research Process," we describe how the first three standards have guided our efforts during the project's first year. The standards for "Application: design" and "Evaluation" will guide our work in the project's second and third years.

Research Design

Our project's goal is to improve patient care by better support of the judgment of BICU clinicians and teams by developing a cognitive aid that assists in decision making and communication. The project's three phases are scheduled to take roughly a year apiece for foundation research, cognitive aid prototype development, and prototype assessment. The first-year goal was to develop a thorough description of individual and team cognition that will provide the basis for cognitive aid prototype development in the second year as well as criteria for prototype assessment in the third year.

The five core team members are experienced in health care field studies using CSE methods and are located remotely from the research site. To manage this, they retained a licensed vocational nurse (LVN) at the site to

TABLE 7.1

Reasonable Scientific Criteria for CSE

CSE Step	Standards
1. Preparation and framing	Clear statements of • Issue or problem • Framing activities outcome • Method, settings, project participant selection rationale
2. Knowledge elicitation	Use of multiple knowledge elicitation (KE) methods Use of interview and observation guides Purposeful sampling of participants and settings Qualified prepared data collectors Quality control protocols (specified format to document data) Manage the dual requirements for rigor and flexibility
3. Analysis and representation	Systematic, purposeful, and documented analysis process Audit trail to connect data elements to findings to design elements Multiple analysis processes and multiple passes thru the data Qualified analysis team members Validity checks on findings Goal-driven selection of qualitative versus quantitative analysis Use of reliability indices
4. Application: design	Iterative design–build–evaluate process Subject matter experts (SMEs) for credibility checks Audit trail to connect data elements, to findings, to design
5. Evaluation	Clear assessment criteria Review evaluation results systematically and purposefully Evaluation methods reflect key cognitive components, behaviors Outcomes reflect cognitive and behavioral issues critical for cognitive work Verify whether the design/changes improve performance

help with the administrative aspects of research team visits. All data collection and human subject consent were carried out under the jurisdiction of the medical center's Institutional Review Board (IRB), which reviewed and approved the research protocol. In advance of the team's first trip to the site, the Co-PI and LVN obtained the consent of health care team members working in the BICU who were willing to participate in the study. Those who declined to participate were excluded from observations and interviews.

Research Site

The research site is a BICU located in a new wing of a federally funded 450-bed tertiary care military academic medical center. The 16-bed unit is widely considered to be one of the best of its kind in the country. Two of the ICU beds are reserved to serve as a postanesthesia care unit (PACU), and another is dedicated to support the center's extracorporeal membrane oxygenation (ECMO) program. Other nearby units support the ICU, including a step-down unit, dedicated burn operating room, and an outpatient clinic.

The typical census averages around 8 patients but has risen to as high as 13 during our study period. This unit's role as a regional tertiary care unit attracts patients who have the most severe affliction from thermal, chemical, mechanical, or electrical burns. It treats patients with burn-like diseases of the skin such as toxic epidermal necrolysis, Stevens–Johnson syndrome, and the autoimmune disorder pemphigus vulgaris. The unit also treats patients with infections or trauma that causes extensive soft tissue damage or loss, such as necrotizing fasciitis, severe degloving injuries, and some war-related trauma. Patient length of stay ranges from days to more than 12 months.

Sample

All clinicians, patients, and patients' friends and family members are potential participants in the study. By the end of the study, we anticipate that over 150 clinicians will be included in the sample. Subjects are recruited through word of mouth in coordination with the BICU medical director and head nurse. Patients in the BICU (or their legal representative) are asked at the start of an observation period to complete a Health Insurance Portability and Accountability Act release before observation or interview. No clinical information collection or recordings are made in the presence of any patient who declines to complete the release. Patient medical data that are necessary to clinical decision making are collected without protected health information and are used only as examples of information that clinicians need to do their work.

Methods

The study of human behavior requires repeated samples to capture its richness, complexity, and variation. No method by itself can account for this complexity. As a result, multiple methods need to be used in order to ensure that the account is valid and as accurate as possible. The research design for this project relies on multiple methods to triangulate data collection and analysis: observation, interviews, and artifact analysis. Comparison of data among all of these sources minimizes the potential bias that a single method may induce.

Observation

In-person observation makes it possible for the research team to witness the phenomena of patient care and team collaboration *in situ*. Informal probe questions enable the researchers to request background and clarifying information in the context of the situation. Observations can be used to study the ways that practitioners perform diagnoses and prepare, launch, monitor, adjust, and complete patient care. The research team performs observations at various times throughout the day and evening to include a range of circumstances and clinicians' responses. Conditions can range from quiet routine to rapid changes. These can happen during the admission or discharge of multiple patients, emergent conditions such as treating rare emergencies like cardiac arrest or burn shock, and common emergencies such as treating postoperative hemodynamic instability.

Observation also includes informal interviews with clinicians as they work in order to learn the bases for their decisions or apparent indecision, motivations, expectations, and preferences that observation alone cannot reveal. Field notes that researchers make during observation provide data for analysis to reveal patterns among and across clinicians. Observations make it possible to describe the ways that individuals and groups cope with complexity and uncertainty. Research team members pay particular attention to heuristics (rules of thumb), and clinicians have developed their expertise and knowledge about individual and system performance, how they use systems such as the electronic health record, mental simulations they perform, and how they assess outcomes. The research team also watches for how the unit members resolve discrepancies and conflicts, negotiate trade-off, evaluate the credibility of data and information from others outside of the unit, and mentor and coach junior members.

During the first visit, team members visited the unit for five weekdays during the day shift (0800–1600). The team scheduled regular observations on the ICU to avoid interfering with clinical work. Subsequent visits to the site also covered evening and night shifts.

Structured Interviews

Cognitive task analysis (CTA) interviews are used to elicit knowledge from clinicians on their background to learn point of view, work activities, information sources on which they rely, and reflections on the challenges they face (Crandall et al. 2006).

Artifact Analysis

Clinicians use cognitive artifacts to capture and share information (Hutchins 2000). These include hard-copy printouts such as sign-out sheets, white marker status boards, and diagnostic and therapeutic equipment displays. They also include personal notes and related items that individuals find helpful, which are not part of the formal information ecology. The research team is collecting de-identified examples of these artifacts that are maintained by and for the group, as well as artifacts that individuals create and use in their work. Both formal and informal artifacts help to understand the inventory of information that the unit develops and uses, which will suggest the content and flow of information that this project's prototype will help to manage.

Research Process

The team began its work by conducting orientation interviews with selected clinicians at the research site. Quality standards described in Table 7.1 that guided our work are shown in italics in the "Preliminary Findings" section. The interviews sought information about the BICU in order to develop an interview guide that would be used to organize data collection efforts during field visits. This enabled the team to develop clear statements of the issues and challenges and the outcome of framing activities. Using these, the team could create the rationale for method, settings, and selection of project participants at the research site. Four one-week data collection visits were conducted at the research site every other month, relying on quality control protocols to document interviews and observations, and cross-check the content of data records. Purposeful sampling of participants and settings ensured validity and reliability of the data that were collected during each visit. Each observation period lasted one week and was followed by a refractory period, during which the investigators reviewed notes, recordings, and artifacts. Data analysis results were also used to revise plans and interview guides for later data collection efforts.

Data Collection

A team of four qualified, prepared data collectors traveled to the site for the first data collection visit. They conferred with the Associate PI (located at the research site) on ICU census and plans for clinical activity. Using multiple KE methods to support findings consistency and comprehensiveness, they conducted CTA interviews to account for each role in the clinical care team. They accompanied the clinical team on daily rounds each morning, which were typically held outside of each patient room. During the trip, the team managed the dual requirements for rigor and flexibility by following interview guides, yet taking the opportunity to shadow participants and ask probe questions when the occasion presented itself. The team collected data firsthand by observing the phenomena that occurred while clinicians provided care in the ICU, using the CSE approach to describe the ICU as a work domain and to account for individual and team cognitive activities. They also collected de-identified examples of computer-based and hard-copy artifacts that the staff use in their daily work.

Rounds were recorded using a handheld video camera to capture team interaction and artifact use and were de-identified using a video-editing software. Recordings were made for future reference on how team members use and share information, including reference to artifacts such as sign-out sheets and task lists. When clinicians interacted directly with the patient, the team used audio recordings to capture how information was shared. No video was taken of the patients. When clinicians had time available, two team members conducted a CTA interview following the interview guide that was developed in the initial six months of the project. If the clinicians were not available during the scheduled team visit, the on-site research nurse would help to organize the interview, and the core team members would participate remotely.

Data Analysis

Data are evaluated using goal-driven selection of qualitative vs. quantitative analysis to extract patterns and themes. The research team gathers for data analysis meetings roughly a month after each data collection visit. The team has experience to detect and elicit patterns through a systematic, purposeful, and documented analysis process. Analysis sessions stimulate insight into what matters in the research setting and why it matters by performing checks on findings credibility, consistency, comprehensiveness, and centrality.

Team members prepare by reviewing the data collected from the most recent visit to ensure that each member has a current accurate recollection. This may also include organizing the data and checking to make sure that they are complete and ready to be analyzed. Members assemble as a group in 2–3 day-long sessions over a week to discover what the data mean by looking for central questions, issues, and themes. For example, the interview guide sought information on how team members manage work flow. Data analysis discussion explored observation notes and interview responses for items related to workflow.

The analysis sessions are intense sense-making exercises that use multiple analysis processes and make multiple passes through the data. Qualified team members use interview notes, observation notes, and artifacts to find patterns and themes in the collected data using reliability indices such as intercoder reliability (when and if they are appropriate). The team also looks for related themes, such as whether there is evidence among the data that show how the clinicians identify and reconcile goal conflicts or resolve agendas that do not agree. Team members suggest themes or patterns that seem to occur in the data. Others challenge, modify, or add to the discussion to ensure validity checks on findings. Team members create diagrams, tables, timelines, and storyboards and use other visualization methods to pose, assemble, and reassemble relationships in order to recognize possible patterns among and across data. During these free-flowing exchanges, new insights rapidly evolve and take the team to a new level of understanding.

Keeping track of the logic trail during these sessions can be a challenge. Maintaining the logical connection from data through analyses matters, because each of the requirements that the analyses eventually produce must have a deliberate link to the data from which they were derived. To keep track of these relationships, the team keeps notes that maintain an audit trail to connect data elements to findings to design elements. Without this structure, it is easy to disregard the data, producing a result that is not a set of findings but rather a collective team impression.

By the end of the analysis sessions, the team has deepened their understanding of what they know about the work setting and what occurs there. They also have a clearer sense of what isn't known yet and needs to be included in the plan for the next site visit. Later in the year, further analysis work will code and analyze all interview and observation data to detect themes and barriers and produce requirements for the prototype.

Limitations

Modest project funding made it necessary to study one site, which limits its reliability. The research team was not available on the unit continuously during the study, making it difficult to observe momentary changes in unit activity such as clinician responses to codes. To mitigate that limitation, the research nurse was available at the research site to collect data in the periods between research team visits.

Preliminary Findings

While the project has only been underway for a brief time, the first data collection and analysis sessions made it possible to describe initial findings that include unit activity, the network of care providers, and information sources on which the clinicians rely. These elements amount to an initial inventory of the work setting that the team can build on during subsequent site visits.

Unit Activity

While many activities occur on the unit through 24 h, Table 7.2 shows the essential events that occur regularly each day. Those who are involved in these activities and the information resources they use to perform them start to flesh out a description of the unit.

Through the evening, the bedside nurse and resident both monitor and occasionally provide medication to the patient assigned to their care. From 6:30 to 8:00 a.m., the residents and medical students examine the patients and prepare for formal multidisciplinary rounds. The Assistant Chief Nurse and oncoming bedside nurses hold a safety huddle. Off-going and oncoming bedside nurses review their patient's condition and conduct a handoff. The ICU Chief Nurse reviews the unit population and resource needs, and the unit dietician reviews patient nutrition plans. At 8:00 a.m., the general rounds begin and can last up to two or more hours depending on a number of factors including unit census, patients' condition, and time pressure. From 8:00 a.m. to 2:00 p.m., patients are showered, receive care for their wounds, or are taken to the nearby operating room procedures such as tissue debridement, skin grafting,

TABLE 7.2

BICU Schematic Timeline—Weekdays

Time	Activity	Participants	Information Resource
0000–0645	Patient monitoring, occasional medication	Bedside nurse; resident	Patient monitors
0630–0800	Patient exam, rounds preparation	Resident, medical student	Sign-out sheet; patient health record (PHR), wound flow, radiology images; patient monitors; bedside nurse, off-going resident
0645–0700	Safety huddle	Assistant Chief Nurse, oncoming bedside nurses	Personal notes
0700–0800	Bedside report and physical assessment	Off-going bedside nurse, oncoming bedside nurse	Patient monitors
0700	ICU audit	Assistant Chief Nurse	Personal notes
0700–0730	Metabolic assessment	Dietitian	Excel file; PHR
0800	Patient rounds	Intensivist, burn surgeon, fellow, resident, bedside nurse, charge nurse, medical student, respiratory therapist, occupational therapist, social worker, dietician, psychiatrist	PHR
0800–1400	Shower, wound care	Bedside nurse, wound care team: RN and LVN	Wound flow
0800–1400	Medications	Bedside nurse	
0800–1400	Surgeries	Burn surgeon, OR team	Shadow charts
~1400	Patient exam	Resident	
1200–1300	Lecture	Staff physician, surgical and medical residents, medical students	
~1500	Afternoon rounds		
1530	Plan for wound care the next day	Charge nurse, wound care coordinator	4T assignments sheet

and reconstructive surgery. The remainder of the day includes a lecture for residents/medical students, the resident examination of his/her patient, brief afternoon rounds to review what has been completed from tasks assigned during morning rounds, and an informal discussion between the wound care team leader and the charge nurse to decide patient plans for the next day.

Network

Patients on this BICU typically need care by a variety of specialists, requiring exceptional planning, coordination, and ability to work together. Table 7.3

TABLE 7.3

BICU Patient and Patient Care Staff Roles

Patient	Bedside Nurse	Patient Family	Attending Intensivist	Burn Surgeon	Licensed Social Worker
Head nurse	Occupational therapist	Respiratory therapist	Resident	Medical student	Clinical nurse specialist
ICU nurse	Psychiatric nurse	Unit clerk	ICU director	Charge nurse	Pharmacist
Staff psychiatric nurse practitioner					

depicts many of the roles that need to collaborate to create and manage a feasible plan for patient care across multiple shifts through the week and the weekend. The roles range from the bedside nurse, who serves as a primary care provider and kind of the gatekeeper for patient care by others, to primary care physicians such as the intensivist and burn surgeon, and care specialists such as the respiratory and occupational therapists, those who care for members of the health care team such as the psychiatric nurse practitioner, managers who assist with planning and oversight, and hospital employees off the BICU such as the pharmacist. In a unit that involves as many team members and specialties as this BICU, it can help to focus on a single most important element of the work domain. In this unit, the bedside nurse is closest to the patient and can serve as a focus of attention for the researcher to understand crucial working relationships. Figure 7.1 represents the 31 working relationships in our data that the bedside nurse maintains in daily practice. Among all of these roles, the bedside nurse interacts most with others on the nursing staff, the patients' family and friends, physicians (including physicians of different levels of training and of different specialties), rehabilitation/occupational therapy technicians, and the clinical lab and blood bank.

Information Resources

Prior work by researchers including Xiao et al. (2001), Wears et al. (2007), Nemeth et al. (2006a), and Bisantz et al. (2010) has described the role of cognitive artifacts (Hutchins 2000) in the health care setting. These artifacts include physical items that are either personal (e.g., a sign-out sheet or note on a scrap of paper) or informal and used by a group (e.g., marker board), as well as electronic information displays that are local (e.g., equipment information display) or distributed (e.g., information system display; electronic medical record). Figure 7.2 depicts many of the artifacts that the staff relies on to perform individual and team cognitive work each day.

Databases and interfaces to manage them include the PHR, outpatient record, blood glucose management, laboratory culture, nurse scheduling, and radiology images. While used in concert, many of these systems are actually

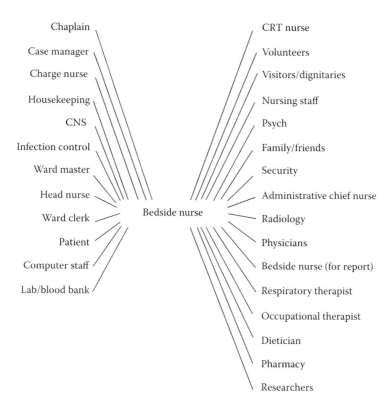

FIGURE 7.1
Initial representation of bedside nurse work relationships. (Copyright © 2013 Applied Research Associates, Inc.)

separate. This separation requires care team members to take extra steps and make temporary hard-copy notes to use and transfer information among systems. Other information resources beyond databases include white boards, a daily wound care plan, vital signs flow list, email/cell phone roster, landline phone roster, resident sign-out sheet, and a charge nurse checklist. The strong emphasis on research at the project site has made it possible for clinicians to develop their own formal electronic information sources in addition to the hard-copy artifacts that may be found at other health care locations. The Wound Flow software program makes it possible to identify the location and condition of tissue injury and skin grafts. An Excel file that the unit dietitian has developed makes it possible to accurately track the quality and amount of nutrition that is crucial for burn patient recovery. The Burn Resuscitation Decision Support software enables the staff to accurately manage fluid resuscitation during the critical 48 h following a significant burn injury. The solution that this project creates will need to bring these various parts of this information ecology (Nemeth et al. 2008) together in order to form a cohesive whole for the unit to use. We expect that using the cognitive aid will enable the unit

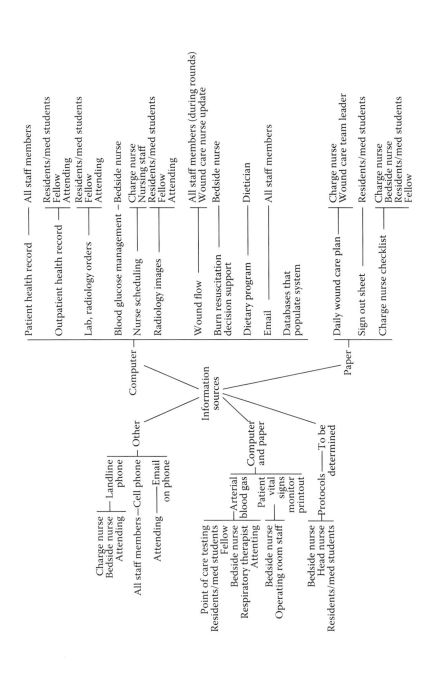

FIGURE 7.2
Information sources that the BICU care providers use. (Copyright © 2013 Applied Research Associates, Inc.)

staff to work together more effectively and efficiently and, as a result, improve patient care effectiveness and outcomes.

Cognitive Work

An initial review of the data indicates that individuals and teams perform a number of macrocognitive (Crandall et al. 2006) activities, which are summarized in Table 7.4. The staff performs *rework* through bridging and work-around strategies to link systems that don't talk to each other in an effort to ensure *information continuity*. For example, the ABG unit is not connected to the database for the electronic PHR. (See Chapter 6 of this book for additional examples, and a proposed model, for tracking ways that information is maintained throughout health care systems.) The dynamic activities on the unit require *negotiation* hourly/by shift/daily among individuals, specialties, and those who have different levels of expertise. *Allocation of resources requires planning and replanning* among and across patients and specialties in *anticipation* of the patient status and needs, as well as how to meet them through preparation and participation in events.

TABLE 7.4

Emergent Themes for Cognitive Work of Burn ICU

Theme	Definition
Rework	Bridging and work-around strategies to link systems that don't talk to each other.
Information continuity	Arterial blood gas (ABG) does/doesn't connect to electronic PHR. An additional volume needs to be created for a very long term care patient.
Negotiation	Among individuals and care specialties, team member levels of knowledge and expertise are dynamic, which requires negotiation by the hour, shift, and day.
Scheduling	Planning and replanning among and across specialties.
Anticipation	Patient status, needs, and how to meet them; preparation and participation in events.
Coordination	Collaboration requires expression of expectations, prioritization, agreement, and recruitment/transfers.
Clarification	Inquiry, sense making, common grounding, to drive down levels of uncertainty and reach an acceptable level of confidence.
Resources	Access, availability, permission, provision, preparation, authority, certification, and use related to equipment, medications, and supplies.
Tasking	Assignment of ICT team members to best match patient needs; based on individual abilities and experience and team needs.
Cross-checking	Identify, confirm, and correct information; problem detection, which may create drag in completing care activities.
Tracking	Account for what needs to be done, whether it has been completed, and what remains to be done.
Gaps	The ability some more experienced team members have to suspect something that is needed is missing.

Collaboration requires the expression of expectations, prioritization, and agreement for staff member recruitment and patient transfers. In order to reach threshold of confidence with which they are comfortable, staff members *clarify* through inquiry, sense making, and seeking common care by reducing uncertainty. Use of *resources* such as equipment depends on its availability as well as permission, provision, preparation, authority, and any required certification to use them. These traits fit what Cook and Woods (2002) have described as the "technical work" in the context of health care. *Tasking* assigns ICU staff members to best match individual abilities/experience and team needs to meet patient needs. Through *cross-checking*, the staff detects problems and identifies, confirms, and corrects information. Their *tracking* efforts account for what needs to be done, whether it has been completed, and what remains to be done. Staff members with the greatest expertise are able to see "gaps," which are, in effect, "what isn't there" but should be.

Challenges

A number of work domain issues shown in Table 7.5 can detract from the time and effort that could be devoted to patient care. Our project team considers each issue from the viewpoint of whether the cognitive aid could help to either mitigate or eliminate them. Nurses fill gaps in the *limited orientation* that residents and float (off unit) nurses receive, which takes time from patient care. Due to *lags in information* timing of information on labs and blood cultures, staff members need to rely on verbal orders (referred to as "on the sly") that are not fully socialized or shared and can result in care delays. *Bedside nurses reconcile conflicts between* patient care needs and technology protocols, guidelines, policy, and regulations. *Procedural drag* results from the need for transcription and work-arounds due to system organizational gaps. The need for clinician *reliance on memory* provides the researcher with a marker for failure, as technology fails to support the needed work. *The long-term story of the patient/big picture is lost*, because trend information and understanding are lost or degraded over a long term of care. *Reliance on verbal exchanges* makes the flow of information porous, brittle, erratically shared, and less reliable. The *authority gradient* between junior and more senior staff members encourages passivity with regard to concerns and impedes sharing. *Common grounding accuracy* suffers from underspecification, requiring confirmation, verification, and clarification. It is not always clear *who has the "Con?"* (has the lead) among specialists during procedures when care quality is high, but no individual takes accountability to assure results. *Timing* issues can result in poor coordination and stale information, such as when a procedure was performed. Without *salience* to bring it to the clinician's attention, important patient information such as "stat" orders is lost in homogenous information displays. Software *usability/access/usefulness* issues result in difficulties in being able to use it, having the knowledge it requires to use it, and being able to enter data accurately.

TABLE 7.5

Emergent Themes of Barriers and Challenges to Effective Care

Issue	Definition
Limited orientation	Residents and float RNs receive limited orientation to the unit. RNs provide orientation, which takes time from patient care.
Lags in information, medications, labs, and blood	Reliance on verbal orders "on the sly" (informally) that are not fully socialized or shared; creates consistent care delays.
Bedside nurse reconciles conflicts	Technology protocol, guidelines, policy, regulations, and patient care needs require choices to be made.
Procedural drag	The need to create work-arounds and bridging tactics to fill the gap between incompatible systems slows down work efficiency.
Reliance on memory as a failure marker	Technology fails to support necessary work, causing clinicians to rely on memory for continuity (e.g., action items not completed by afternoon rounds not carried through to the next day).
Story of the patient/ big picture is lost	Incremental views of patient status are not synthesized into a whole picture; particular concern for patients in BICU for extended periods.
Reliance on verbal exchanges	Information flow is porous, brittle, not shared, or reliable.
Authority gradient	Encourages passivity with respect to expressing concerns.
Common grounding accuracy	Under specification, needs for confirmation, verification, clarification all affect ability of clinicians to develop consensus.
Action/who has the "Con?"	Numerous well-qualified clinical specialties collaborate but lack of clarity regarding who is leading a particular procedure (e.g., ECMO).
Timing	Lack of synchrony can result in stale information (e.g., when the procedure was performed).
Salience	Great deal of information that is presented homogenously. Information that is most relevant is difficult to find (e.g., "Stat" orders are not evident).
Usability/access/ usefulness	Systems cannot be used without requisite operator knowledge, certain access requirements.
Organizational issues = drag	Compliance with administrative reminders detracts from patient care.

Compliance with *organizational issues* such as administrative reminders creates drag for clinician efficiency.

Discussion

The ICU Work Setting

ICU patients present clinical teams with unique challenges and complex combinations of life-threatening injuries and illnesses. Care for this patient population is necessarily multidisciplinary and includes many specialties.

Care providers across these clinical areas must collaborate to develop treatment plans, assess progress, and refine or change treatment plans and modes.

Clinician decisions are only as good as the information that is available when they are made. The daily work on the unit requires representations that serve as a map of the ever-changing environment of work that must be successfully navigated. Clinical teams that care for ICU patients in the military health care system encounter these challenges as they make diagnostic and therapeutic decisions and share them with colleagues. Decision-making difficulty increases as the number of patients and the severity of their conditions increase. Complexity grows as the number of care providers seeks to make their own unique contribution to a patient's care.

Patient care activities rely on the acquisition, portrayal, and analysis of therapeutic and diagnostic information from many sources. This creates a complex work setting that is composed of multiple independent agents. All interact in various ways according to inconsistent rules in an attempt to adapt to changing conditions. Because of this, the organization's outcomes are unpredictable, but they often follow predictable patterns (Plsek and Greenhalgh 2001).

Other ethnographic studies also revealed insights into acute care settings. For example, Fackler et al. (2009) used CTA to identify cognitive aspects of critical care practice in two academic ICUs and identified broad categories of cognitive activity: pattern recognition; uncertainty management; strategic vs. tactical thinking; team coordination and maintenance of common ground; and creation and transfer of meaning through stories. Anders et al. (2012) used a simulator-based experiment to evaluate ICU nurses' ability to detect patient changes using an integrated graphical information display (IGID) compared with a conventional electronic chart-style ICU patient information display. The study found that the 32 ICU nurse samples reported more important physiological information with the novel IGID compared with the tabular display and concluded that information displays should accommodate the diversity of those who are intended to use it.

Novak et al. (2012) found that medication administration intersects with other organizational routines, and IT-enabled changes to one routine lead to unintended consequences in its intersection with others. Introducing IT can be improved by nurses who provide technology-use mediation before and after the rollout of a new health IT system. Their efforts can help others to better understand the relationship between IT introduction and changes in routines.

In addition to operational complexity, our research into reporting health care adverse events using CSE methods (Nemeth et al. 2006b) has also revealed technical, social, political, and legal forces. Each influences acute care settings such as the ICU, which are typically uncertain, interrupt driven, saturated, and contingent.

Uncertain: Clinicians must treat widely varying patient populations. Time pressure can force clinicians to make decisions based on information that can be insufficient or ambiguous. Field studies using CSE methods can discover initiatives that clinicians have developed to minimize uncertainty.

Interrupt driven: Interruptions create breaks in clinicians' task-oriented work (Chisholm et al. 2000), and when they occur during diagnosis and treatment, they can degrade or defeat attempts to treat patients. Work domain study using CSE methods can identify gaps in care continuity, detect how clinicians allocate limited attention reserves, and produce tools such as cognitive artifacts that maximize patient care opportunities.

Saturated: Facilities and staffs typically run at or near capacity. With little margin of time or resources to spare, clinicians have to develop strategies to cope with variations in care demand. Work domain studies using CSE can reveal discontinuities that exist in the match between resources and demand, such as late shifts, and unexpected surges in care demand.

Contingent: The process of care depends on the patient, including presenting symptoms, documentation of history, response to therapy, expected trajectory of treatment, compliance, and more. CSE methods can be used to discover how care providers create, monitor, and adjust multiple contingencies in order to achieve as satisfactory and expedient an outcome as possible for patients.

In addition, distraction, complexity, remote influences, and consideration make health care human subjects research a particular challenge.

Distraction: Many activities are performed by a variety of clinicians in the vicinity of each other. This makes it easy to be distracted by phenomena that are not necessarily key features of the work domain.

Complexity: Acute care settings have many complex activities that occur at the same time. This is particularly true in an ICU.

Remote influences: Care team members can be distributed across various locations and across time. Not all activity that matters occurs within view or in the immediate recall of those whom the researcher interviews.

Consideration: Patients in the BICU are typically fragile as a result of some trauma. This calls for the researcher to have an adequate sensitivity to care providers, patients, and the patient's family members.

All of these influences form the context in which clinicians perform their cognitive work. The CSE approach makes it possible to describe the domain and individual and team activity in it to transform findings into requirements that serve as the basis for a prototype cognitive aid.

Communication among Care Team Members

Team communication creates, and is created by, the work context. CSE can be used to reveal the context and worker behaviors that lead to understanding communication needs and how to support them. This contrasts with the more traditional information engineering approach that assumes that understanding comes simply from the faithful uninterrupted transmission of data (Feldman and March 1981; Stohl and Redding 1987). Care provider expectations differ on communication content, form, relevance, and value of its completeness.

Interventions based on CSE methods can benefit team communication. For example, Grome et al. (2009) found that co-creative development workshop helped surgical team representatives to create and adapt preoperative briefing content and structure, as well as measures to assess the briefing's effect on teamwork, communication, and patient safety.

Nemeth and Cook (2013) used CSE to identify barriers that can erode the quality and reliability of health care communication that this project addresses.

Difficulties in communication. Health care and the information that is needed to provide it are typically complex and demand accuracy in order to avoid misinterpretation.

Confusion of responsibility. Interwoven relationships among care providers, units, departments, and institutions can result in confusion over who is responsible for a patient's care.

Lack of, or variable availability of, good information resources. Even with sophisticated information technology available, system failure or incompatibility can result in images and reports being mislabeled, misunderstood, swapped, late, misidentified, or unavailable.

Work environment pressures. Care provider efforts to cope with workload demands and time pressure can result in a kind of "shorthand" that edits information in order to be efficient.

Lack of standards or training. Clinical specialties and institutions can vary in the way they go about practices such as handoffs, resulting in the potential for misperception.

Aptitude. Patients and family members may find it hard to understand the information that is conveyed through written, verbal, and graphic health care communication.

Attention. Understanding and context are essential to effective communication. Simple transmission (e.g., a "data dump") does not guarantee that others understand what is provided or can correctly put it into context.

Attitude. Clinician empathy may yield a number of benefits, including patients reporting more about their symptoms and concerns, increased physician diagnostic accuracy, patients receiving more illness-specific information, increased patient participation and education, increased patient compliance and satisfaction, greater patient enablement, and reduced patient emotional distress.

Reader et al. (2008) found that team structure and individual roles and stature have significant effect on ICU communication, and a difference in status appears to influence how communication is perceived. The "authority gradient" barrier mentioned in Table 7.5 may be related to this issue.

   Through the use of CSE, the cognitive aid that this project produces will need to help the ICU staff to overcome these potential barriers.

The Role of CSE

The use of CSE methods makes it possible for the researcher to "get in" at the right level of detail. Too general a study will miss the nuances and refinements that clinicians create in order to make their work possible. Too detailed a study may collect great amounts of data but will also miss the broader patterns that make insight possible. Studies of such a complex domain require repeated visits in order to reveal the deeper aspects of what occurs. These are what have been referred to as the "messy details" of technical work (Nemeth et al. 2004). The researcher needs to learn about real-world settings that involve the organized activities of daily life (Garfinkel 1967). Real-world settings are stubborn, though, and do not easily reveal themselves (Blumer 1969).

   Research can be basic (a search for general principles), applied (adapting general findings to classes of problems), or clinical (related to specific cases). Most design research is clinical because time and budget allow for little else (Friedman 2000). CSE methods can be used to negotiate the gap between applied and clinical research.

CSE in Health Care

Recent work on collaboration has produced distributed cognition and joint cognitive system models that can be used to better understand health care as a collective enterprise. The use of CSE to identify and describe all ICU elements, including clinicians, information, and artifacts, can identify system gaps. Addressing gaps can lead to authentic improvement in performance and outcomes. For this reason, CSE is particularly well suited to the discovery of phenomena in complex real-world settings.

   Distributed cognition (Hutchins 1995) is the interaction of individuals, artifacts, and the environment. Practitioners must rely on this to prevent the formation of gaps in the continuity of care (Cook et al. 2000). This includes transfers between departments, work-cycle shift changes, and information exchanges among professionals from different fields of practice. Clinicians in an ICU comprise a joint cognitive system that can modify its behavior and decision making on the basis of experience in order to maintain order (Hollnagel and Woods 1983). The daily work of the clinician requires representations that serve as a map of the ever-changing environment of work that must be successfully navigated (Rasmussen et al. 1994). Individual elements of information vary enormously in the length of time that they are reliable, and their value depends on their context. What is represented and how it is represented should depend on the cognitive work it is intended to support. Furthermore, the partial and overlapping interaction among clinical specialties in the ICU lends itself to additional gaps in care continuity and the misadventures that can result.

Validity

Nemeth et al. (2011) recommended four ways to verify whether results from qualitative studies such as this ICU research project are valid. Findings must be credible, consistent, comprehensive, and central.

Credible. Do findings "ring true" to SMEs and others who work in the domain?

Consistent. Do findings replicate across interviews and across incidents?

Comprehensive. How general are the findings? To what range of tasks and settings do they apply? Can boundaries be identified, and can those limitations be stated?

Central. Do findings speak to cognitive issues that *matter* for performance based on SME judgments, research literature, and other sources?

Studies that meet these criteria are more likely to pass validity tests when solutions are evaluated.

Aspects of Resilience

Knowledge gained through the use of CSE about the nature of work as it is actually done can help to contribute to the system's ability to adapt when confronted with unforeseen challenges—to be more *resilient* (Hollnagel et al. 2006). Recent writing in resilience engineering has identified a number of system characteristics that contribute to system resilience. This knowledge can improve their ability to operate despite significant challenges such as changes in the type, rate, and volume of care. Three characteristics that CSE can assist include being self-aware, the ability to identify and apply resources, and the ability to adapt to surprise.

Self-Aware

The "cottage industry structure of the national healthcare delivery system" results in "disconnected silos of function and specialization." (Reid et al. 2005, pp. 12–13) Acute and ambulatory care patients require coordinated care that is provided by multiple distributed care providers. Their care also calls for the coordination and integration of many functions and specialized areas of knowledge over time. Yet connectivity, integrated care, and coordination are inadequate nationwide at all stages of illness treatment. An estimated 60 million patients in the United States suffer from two or more chronic conditions and are particularly affected by the disconnection among clinical care specialties. The ability to reveal the nature of work domains by using CSE can start to mitigate this significant and widespread issue.

Able to Identify and Apply Resources

Skills, supplies, equipment, and facilities are routinely assembled to perform each procedure. CSE can be used to document work processes and what influences them. That can lead to insight into how these configurations are developed and managed, what goes well, and where misadventures can occur.

Able to Adapt to Surprise

We have shown in prior publications (Nemeth et al. 2007; Cook and Nemeth 2010) how health care organizations respond to events, particularly misadventures. More often than not, the response attempts to isolate the cause and declare that it will not happen again. These efforts stop the exposure to risk. However, they also stop the learning that can inform us how systems have difficulty adapting. The use of CSE makes understanding what goes right, and what occasionally does not, a routine learning process that can improve the ability to adapt.

Summary

We need to learn what people actually do in health care teams and how to design work processes and systems based on that knowledge. This calls for an approach that reveals the true nature of work as it is actually done, not as it is intended to be done. CSE serves that purpose well.

Early data collection and analysis activity in our BICU research have identified the network of those who care for patients, the information sources they use, and the flow of patient care activity. Continued visits are expected to deepen the understanding of interrelationships among clinicians, how they address and resolve conflicts such as different agendas, the information sources and their use, and cognitive activities for each of the clinical specialties and roles. Results from this first year of study will be used to develop requirements for decisions that clinicians make. Requirements and use cases will provide the basis for a prototype to be developed and evaluated in the project's second and third years.

The well-designed valid cognitive artifact that results from our use of CSE is intended to support individual and team cognitive work, which is expected to improve the reliability and efficiency of clinical care for patients.

Acknowledgments

The authors are grateful to team members Cynthia Dominguez, PhD, Greg Rule, and Dianne Hancock for their collaboration, as well as the LTC Elizabeth Mann-Salinas, LTC Kevin Chung, and clinicians at the study's research site for their generous support. This work is supported by the US Army Medical Research and Material Command under Contract No. W81XWH-12-C-0126. The views, opinions, and/or findings contained in this report are those of the author(s) and should not be construed as an official Department of the Army position, policy, or decision unless so designated by other documentation. In the conduct of research where humans are the subjects, the investigator(s) adhered to the policies regarding the protection of human subjects as prescribed by Code of Federal Regulations Title 45, Volume 1, Part 46; Title 32, Chapter 1, Part 219; and Title 21, Chapter 1, Part 50 (Protection of Human Subjects).

References

Anders, S., Albert, R., Miller, A., Weinger, M.B., Doig, A.K., Behrens, M. and Agutter, J. (2012). Evaluation of an integrated graphical display to promote acute change detection in ICU patients. *International Journal of Medical Informatics*, 81, 12, 842–851.

Bisantz, A.M., Pennathur, P.R., Guarrera, T.K., Fairbanks, R.J., Perry, S.J., Zwemer, F.L. and Wears, R.L. (2010). Emergency department status boards: A case study in information systems transition. *Journal of Cognitive Engineering and Decision Making*, 4, 1, 39–68.

Blumer, H. (1969). *Symbolic Interactionism*. Berkeley, CA: University of California Press.

Chisholm, C.D., Collison, E.K., Nelson, D.R. and Cordell, W.H. (2000). Emergency department workplace interruptions: Are emergency physicians "interrupt-driven" and "multitasking"? *Academic Emergency Medicine*, 7, 11, 1239–1243.

Cook, R. and Nemeth, C. (2010). Those found responsible have been sacked: Some observations on the usefulness of error. *Cognition, Technology and Work*, 12, 87–93.

Cook, R. and Woods, D. (2002). Nine steps to move forward from error. *Cognition, Technology, and Work*, 4, 137–144.

Cook, R.I., Render, M. and Woods, D. (2000). Gaps in the continuity of care and progress on patient safety. *British Medical Journal*, 320, 7237, 791–794.

Crandall, B., Klein, G. and Hoffman, R.R. (2006). *Working Minds: A Practitioner's Guide to Cognitive Task Analysis*. Cambridge, MA: The MIT Press.

Fackler, J.C., Watts, C., Grome, A., Miller, T., Crandall, B. and Pronovost, P. (2009). Critical care physician cognitive task analysis: An exploratory study. *Critical Care*, 13, 2.

Feldman, M. and March, J. (1981). Information as signal and symbol. *Administrative Science Quarterly*, 26, 2, 171–186.

France, D.J. and Levin, S. (2006). System complexity as a measure of safe capacity for the emergency department. *Academic Emergency Medicine*, 13, 1212–1219.

Friedman, K. (2000). Creating design knowledge: From research into practice. *IDATER 2000*. Leicestershire, UK: Loughborough University.

Garfinkel, H. (1967). *Studies in Ethnomethodology*. Upper Saddle River, NJ: Prentice-Hall.

Grome, A., Crandall, B., Brown, J.P., Sanford-Ring, S. and Douglas, S. (2009). Patient safety in the operating room: Improving team coordination and communication. *Proceedings of the Interservice/Industry Training, Simulation, and Education Conference*. Orlando, FL.

Hollnagel, E. and Woods, D.D. (1983). Cognitive systems engineering: New wine in new bottles. *International Journal of Man-Machine Studies*, 18, 6, 583–600.

Hollnagel, E., Woods, D.D. and Leveson, N. (Eds.) (2006). *Resilience Engineering: Concepts and Precepts*. Aldershot, UK: Ashgate Publishing.

Hutchins, E. (1995). Cognition in the wild. Cambridge, MA: The MIT Press.

Hutchins, E. (2000). *Cognition in the Wild*. Cambridge, MA: MIT Press.

Hutchins, E. (2002). Cognitive artifacts. Retrieved on July 7, 2002 from the MIT COGNET website: http://cognet.mit.edu/MITECS/Entry/hutchins.

Kyriacou, D.N., Ricketts, V., Dyne, P.L., McCollough, M.D. and Talan, D.A. (1999). A 5-year time study analysis of emergency department care efficiency. *Annals of Emergency Medicine*, 34, 3, 326–335.

Levin, S., France, D.J., Hemphill, R., Jones, I., Chen, K.Y., Rickard, D., Makowski, R. and Aronsky, D. (2006). Tracking workload in the emergency department. *Human Factors*, 48, 3, 526–539.

Militello, L., Dominguez, C., Lintern, G. and Klein, G. (2010). The role of cognitive systems engineering in the systems engineering design process. *Systems Engineering*, 13, 3, 261–273.

Nemeth, C. and Cook, R.I. (2013). Improving team communication for better health behavior. In L. Martin and D. Matteo (Eds.), *The Oxford Handbook of Healthcare Communication, Behavior Change and Treatment Adherence*. Oxford Library of Psychology. New York: Oxford University Press.

Nemeth, C., Cook, R. and Woods, D. (2004). The messy details: Insights from technical work in healthcare. *IEEE Transactions on Systems, Man and Cybernetics-Part A*, 34, 6, 689–692.

Nemeth, C., O'Connor, M., Klock, P.A. and Cook, R.I. (2006a). Discovering healthcare cognition: The use of cognitive artifacts to reveal cognitive work. In Lipshitz, R. (Ed.), Special Issue on Naturalistic Decision Making. *Organization Studies*, 27, 7, 1011–1035.

Nemeth, C., Dierks, M., Patterson, E., Donchin, Y., Crowley, J., McNee, S., Powell, T. and Cook, R.I. (2006b). Learning from investigation. *Proceedings of the Human Factors* and *Ergonomics Society Annual Meeting* (pp. 914–917). San Francisco.

Nemeth, C., Dominguez, C., Grome, A., Crandall, B., Wiggins, S. and O'Connor, M. (2011). Setting the bar: Performance standards in naturalistic decision making research. *10th International Conference on Naturalistic Decision Making (NDM2011)*. Orlando, FL.

Nemeth, C., Nunnally, M., O'Connor, M., Brandwijk, M., Kowalsky, J. and Cook, R. (2007). Regularly irregular: How groups reconcile cross-cutting agendas in healthcare. *Cognition, Technology and Work*, 9, 3, 139–148.

Nemeth, C., O'Connor, M. and Cook, R. (2008). The infusion device as a source of resilience. In C. Nemeth, E. Hollnagel and S. Dekker (Eds.), *Preparation and*

Restoration. Resilience Engineering Perspectives, vol. 2. Farnham, UK: Ashgate Publishing.

Novak, L., Brooks, J., Gadd, C., Anders, S., and Lorenzi, N. (2012). Mediating the intersections of organizational routines during the introduction of a health IT system. *European Journal of Information Systems*, 21, 5, 1–32. doi:10.1057/ejis.2012.2.

Plews-Organ, M.L., Nadkarni, M., Forren, S. et al. (2004). Patient safety in the ambulatory care setting: A clinician-based approach. *Journal of General Internal Medicine*, 19, 7, 719–725.

Plsek, P.E. and Greenhalgh, T. (2001). Complexity science: The challenge of complexity in health care. *British Medical Journal*, 323, 7313, 625–628.

Rasmussen, J., Pejtersen, A.M. and Goodstein, L.P. (1994). *Cognitive Systems Engineering*. New York: John Wiley & Sons, Inc.

Reader, T., Flin, R. and Cuthbertson, B. (2008). Factors affecting team communication in the intensive care unit (ICU). In C. Nemeth (Ed.), *Improving Healthcare Team Communication* (pp. 117–133). Farnham, UK: Ashgate Publishing.

Reid, P.R., Compton, W.D., Grossman, J.H. and Fanjiang, G. (Eds.) (2005). *Building a Better Delivery System: A New Engineering/Health Care Partnership*. Washington, DC: The National Academies Press.

Roth, E.M., Patterson, E.S. and Mumaw, R.J. (2002). Cognitive engineering: Issues in user-centered system design. In J.J. Marciniak (Ed.), *Encyclopedia of Software Engineering* (2nd Ed., pp. 163–179). New York: Wiley-Interscience, John Wiley & Sons.

Shojania, K.G., Duncan, B.W., McDonald, K.M., Wachter, R.M. and Markowitz, A.J. (2001). Making health care safer: A critical analysis of patient safety practices. *Evidence Report/Technology Assessment*, 43, i–x, 1–668.

Stohl, C. and Redding, W.C. (1987). Messages and message exchange processes. In F. Jablin, L. Putnam, K. Roberts and L. Porter (Eds.), *The Handbook of Organizational Communication* (pp. 451–502). Beverly Hills, CA: Sage.

Wears, R. and Nemeth, C. (2007). Replacing hindsight with insight: Toward a better understanding of diagnostic failures. *Academic Emergency Medicine*, 49, 2, 206–209.

Wears, R., Perry, S., Wilson, S., Galliers, J. and Fone, J. (2007). Emergency department status boards: User-evolved artefacts for inter- and intra-group coordination. *Cognition, Technology, and Work*, 9, 3, 167–170.

Woods, D. and Roth, E. (1988). Cognitive systems engineering. In M. Helander (Ed.), *Handbook of Human-Computer Interaction* (pp. 3–43). Amsterdam: North-Holland.

Xiao, Y., Lasome, C., Moss, J., Mackenzie, C.F. and Farsi, S. (2001). Cognitive properties of a whiteboard. In W. Prinz, M. Jarke, Y. Rogers, K. Schmidt and V. Wulf (Eds.), *Proceedings of the Seventh European Conference on Computer-Supported Cooperative Work* (pp. 16–20). Bonn: Kluwer Academic Publishers.

8

Matching Cognitive Aids and the "Real Work" of Health Care in Support of Surgical Microsystem Teamwork

Sarah Henrickson Parker and Shawna J. Perry

CONTENTS

Why Look at Work?

Despite the complexity of sociotechnical systems such as health care, the critical embedded element—human beings—is often minimized or ignored during the design phases of improvement efforts. The human is frequently considered only after an implementation has not gone well, while the relationships of workers to one another within the work system are rarely acknowledged. This stance is based on outdated assumptions that workers perform in "unsystematic and irrational ways" and that designers are "clumsy or even stupid" (Vicente 1999, p. xi). Cognitive work analysis (CWA) examines existing work systems and develops recommendations for designing for effective human involvement in the origination and nature of work. CWA views individuals embedded within a work system as "actors" rather than "users," thus including the human as an influential member of each work system (Vicente 1999). Rasmussen and Vicente describe CWA as a method for truly understanding the influence of the work system and the relationship of its components (i.e., tools, technology, tasks, organization, people, environment, training, etc.) on work performance. Vicente et al. charge that researchers have failed to acknowledge that the system actually shapes worker behaviors, and as such, departures from expected behavior are generally the result of workers being clever, rational, and, more importantly, competent in their current work setting. These features of the "human factor" are imperative for the continuation and optimization of work in any complex system and, as such, must be designed for and optimized within their work system.

Work Systems in Health Care

The concept of health care delivery as a form of work being done as part of a "system" rather than the culmination of individual efforts is quite new to the institution of medicine (Reason 1995, 1997; de Leval et al. 2000; Shortell and Singer 2008). Adoption of systems concepts has been difficult as it requires a major shift in perspective from the traditional "name, blame, shame, and train" mentality that is common in health care to a systems point of view. In a systems point of view, human error is accepted as the *result* of a combination of work system factors, one of which is the individual, rather than the exclusive failing of any individual actor, a fundamental tenet within clinical care (Vincent 2003; Wears 2003; Wiegmann et al. 2010).

Because of the significance of this shift, health care has been struggling for over a decade with the cultural and behavioral changes necessary for a paradigm change. A systems approach requires better understanding of existing microsystems, degrees of coupling between them, and system vulnerabilities. This is a very tall order for the work system of health care, which has a high degree of complexity and risk, the majority of which is opaque to those within the system (Behara et al. 2005; Wears et al. 2005).

The domain of surgery has been particularly problematic as it tries to reconcile a systems approach with current perspectives of clinical work (Wiegmann et al. 2010). Traditionally, surgeons have been taught that patient outcomes are the result of only two variables: the surgeon's skill and the patient's risk factors associated with comorbidities (American College of Surgeons 2004; Vincent et al. 2004). In fact, according to the American College of Surgeons' Statement of Principles, "the surgeon is personally responsible for the patient's welfare throughout the operation." This focus exists despite obvious interdependencies between specialties and skill sets present for every surgery (e.g., surgeons, anesthesiologists, nurses, and technicians) and other system factors, such as technology, equipment, and the physical environment. The opportunity to better understand work and work design is critical for improving safety in high-risk health care settings such as the OR.

Focusing on cardiac surgery as a sample health care microsystem, this chapter will discuss this high-risk and very complex sociotechnical work system as well as components and relationships that are difficult to quantify but have tremendous impact on system performance. We will particularly focus on research examining work system factors that influence surgical team performance and communication during cardiac surgery. We will also discuss the development of several interventions to improve teamwork and communication in this microsystem that would be difficult to design without an understanding of system factors and influences. The goal of this chapter is not to demonstrate the development of a "one-size-fits-all" safety improvement but rather to discuss multiple usable and scalable methods to investigate work system problems (no matter the size). The endpoint therein is to highlight the importance of creating solutions that are uniquely suited to specific health care environments and not to health care organizations as a whole.

Setting the Stage: Overview of the Surgical Work Environment

Surgery is performed for the purpose of structurally altering the human body by incision or destruction of tissues and is part of the practice of medicine (Grill 2012). Major surgery, such as cardiac surgery, is delineated by "the requirement for general anesthesia and respiratory support with an associated risk of severe hemorrhage and conditions in which the patient's life is at risk" (Pilcher 1917, p. 799). Surgery by its very nature requires a form of team work (e.g., surgeon to operate, anesthesiologist to sedate the patient), all assembled simultaneously to perform an invasive procedure. Using the Systems Engineering Initiative for Patient Safety (SEIPS) model of work systems (Carayon et al. 2006; see Figure 8.1) in health care, we will set the stage for where and how this type of work occurs.

Based on this model, the "person" or workers involved in any surgical operation are referred to as the OR surgical team. The OR team consists of multiple subteams that work simultaneously to facilitate the work and perform needed tasks, with an emphasis on clinical performance and patient safety. The primary subteams participating in any operation are nursing,

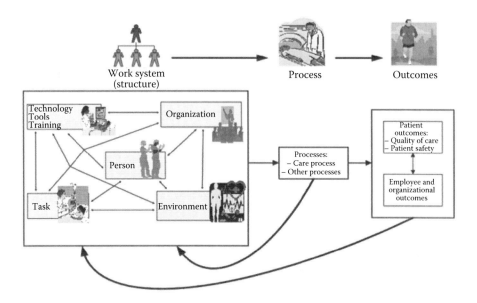

FIGURE 8.1
SEIPS model for work systems in health care. (From Carayon, P. et al., *Quality and Safety in Health Care,* 15(supp 1), 50–58, 2006.)

surgery, and anesthesia (Undre et al. 2006). The individual(s) are defined as teams because they interface directly and interdependently in response to environmental contingencies toward the accomplishment of collective goals (Mathieu et al. 2001, p. 290). Team members may be of disparate expertise, may utilize multiple loci of information, and may need to share multiple types and levels of information in multiple forms both implicitly and explicitly. These workers may or may not be acquainted with each other prior to being assigned to a team (or arriving to the operating room [OR]), and depending upon staffing needs, all but the surgeon may be removed and replaced at any time during the operation.

Each surgical subteam is responsible for very specific aspects of the work necessary for an operation to occur, and each work goal requires many tasks and subtasks. For instance, the anesthesiologist must obtain intravenous access, ensure adequate levels of sedation, and appropriately manage the patient's airway, breathing, and vital signs; the surgeon must perform the surgical procedure, which often requires adjustments and improvisation secondary to unexpected anatomic and pathologic findings; the scrub nurse must support the surgeon by managing instruments and handing them to the surgeon while also monitoring and ensuring sterility; and the circulating nurse must manage all nonsterile activities related to the surgery (blood and tissue samples, pagers, equipment, traffic in the room, etc.), serve as a patient advocate, and anticipate the needs of those directly participating in the surgery, especially the unexpected. Also included in a unique capacity is the patient, whose comorbidities (e.g., whether they are diabetic or have a bleeding disorder) and anatomical anomalies present additional complexities and increased risk of surprise events. With the increasing penetration of health information technology into clinical care, the tools, equipment, and technology used to conduct any surgery have grown remarkably in number and in size (i.e., the robotic da Vinci Surgical System). As a result, there are varying degrees of complexity in an OR (a scalpel vs. an intravenous fluid pump) as well as in worker expertise and the need for integration with other equipment. The resulting variation has major implications for problem solving in the face of developing failure.

The environment of the OR is also an important contributor to the effectiveness of the work of operating on a patient. The built environment is highly variable in size, shape, and configuration, with the assignment of cases often not taking into consideration the number of workers or the amount and size of equipment required. The ambient environment in the OR is kept cold in an attempt to manage infection risk, and it is not ergonomically supportive of the workers (for instance, the surgeon may be standing for several hours on a hard tile floor, or computer-generated x-ray images may only be able to be projected on screens behind the surgeon). The amount of noise present during an operation is highly variable as well, ranging from calm and quiet to tense, loud, and hurried with frequent interruptions.

The layout of an OR to perform the work of an operation generates a significant amount for variation based on the type of surgery being done (e.g., head and neck surgery with a laser will require the patient to be positioned differently than for an abdominal surgery). Each surgeon typically has their own set of equipment or preferences for certain equipment, which is listed on a "preference card" maintained by the administrative staff of the surgical department. The nursing subteam is usually expected to set up the OR per each surgeon's preference, with an emphasis on the positioning of the patient and the surgeon's proximity to the scrub nurse for surgical instruments. The equipment brought into the room is dictated by the type of procedure scheduled and anticipatory decisions by the surgeon (e.g., if the surgeon anticipates large blood loss, a cell saver machine may be requested). The result is a wide range of equipment being present, ranging from surgical hand instruments to anesthesia machines and equipment for patient monitoring to fluoroscopy (x-ray machines) and laparoscopic equipment, among many others.

The major components of any operation can be divided into phases: (1) preoperative preparation (where the patient is assessed and prepared for the OR); (2) movement to the OR and anesthesia administration; (3) the operation or intraoperative phase; (4) postoperative anesthesia recovery (where the patient awakens and initial pain is managed in the OR); and (5) recovery in the intensive care unit or postanesthesia care unit (where the patient is moved to a recovery area for continued monitoring). For the purposes of this chapter, we will focus on the work in the preoperative and intraoperative phases.

No matter how many team members, what equipment, how ill a patient is, or the type of operation, there are inviolate consistencies that are enforced in the OR. First, there are designated areas and degrees of sterility, with some areas requiring surgical clothing (e.g., scrubs, surgical caps, shoe covers) to enter. Within the OR itself, specific sections are deemed for sterile access only and limited to individuals wearing sterile gowns, masks, and gloves, and include at a minimum the operating table, the surgical instrument table, and the patient. Enforcement of these boundaries is strict and deemed enforceable by all workers in surgical areas no matter their skill level. Once the staff have "gowned and gloved" in order to have access to sterile areas of the OR, they may interact with any sterile equipment or touch the patient who at this point has been anesthetized and covered with sterile sheets and materials.

Communication within this complex work environment is highly variable and often limited to within the subsystem teams participating in the operation and even then through various modalities (Henrickson Parker et al. 2012). Presurgical and intrasurgical data related to the conduct of surgery reside in multiple locations such as the medical chart, the minds of members of the surgical team, the minds of the nursing team, the patient, the surgeons' preference card, and the mind of the anesthesiologist. This information is typically shared on a "need-to-know" basis, meaning it is shared beyond the individual or their subsystem should the work change such that there is a necessity to share across teams. For example, the anesthesiologist may note that there is potential for difficulty intubating the patient, which he/she shares with a member of the anesthesia subteam; but this would not typically be shared with the rest of the subteam unless a life-threatening situation develops, or the nursing circulator knows that there will be a shortage of nurses for the operation approximately 2 h after the initial incision and does not share this widely, or the surgeon could be anticipating a difficult surgical dissection but only informs his/her surgical assistant of his/her concerns. Such behavior is not a deliberate malicious activity but rather a function of siloing of the subteams resulting in information exchange often being isolated to the subteam (Edmondson 2003; Lingard et al. 2004). Communication patterns of surgeons reveal that they tend to lead surgical trainees during the intraoperative period, and while they feel responsible for the task of operating, they do not feel that they are the leader of the entire OR team participating in the procedure (Henrickson Parker et al. 2012). Determining which information is important to share and a method for sharing that information is therefore difficult.

Against this general backdrop of the nature of work within a surgical subsystem, we will focus on a specific microsystem: cardiac surgery. This microsystem consists of a large team of expert workers divided into a number of subteams, some of which are nontraditional within the surgery. Cardiac surgery involves very intricate high-risk procedures and is inundated with a highly complex technology, making it an ideal sociotechnical environment in which to examine team performance and communication.

Team Work within the Cardiac Surgery Microsystem

Marked variation found across surgical work exists based not only upon the nature of the medical problem being addressed by a surgical procedure (e.g., an amputation is different from a face lift) but also by the intrinsic properties of the patient (adult or pediatric) and the biology of the work such as the tissue being operated on. Within cardiac surgery, there is a large amount of instrumentation and highly intricate procedures as the veins and arteries are typically very fragile. If valve replacements, stents, or other implantable devices are used, the surgeon may be forced to wait to choose the particular implant until in the midst of the operation, as they may only have the best understanding of the extent of disease once the chest is opened. The most unique to cardiac surgery, however, is the necessity to operate on the heart, which in many cases will need to cease beating for long periods during the surgery, often lasting more than 6 h. This results in the addition of more workers to the surgical team, thus requiring new communication structures and patterns, and bringing even greater complexity to this particular clinical discipline.

In a typical cardiac OR, subteams are further delineated from the traditional subteams discussed earlier (surgical, nursing, and anesthesia) into additional teams determined by their relationship to the established zones of sterility (the center square in Figure 8.2). There is a scrubbed surgical

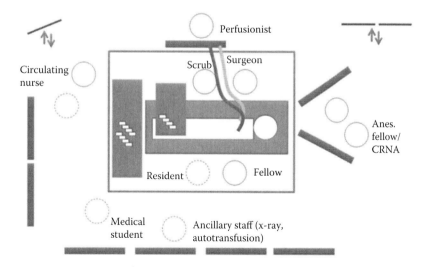

FIGURE 8.2
Typical cardiac OR team and equipment layout.

team and an unscrubbed team. The scrubbed cardiac surgical team, which is allowed only within the zone of highest sterility in proximity to the patient, includes surgical and nursing personnel—the surgeon, scrub technician, or nurse, and in an academic setting, a surgical resident or a more advanced surgical fellow, and may include a surgical assistant. The unscrubbed team is composed of not only anesthesia and nursing but also additional workers with specialized skill sets for managing technology to support the patient during the operation once the heart has been stopped, such as a perfusionist. None of these individuals are allowed within the zone of highest sterility at any time (Figure 8.2).

The perfusionist and monitoring technicians support the placement onto and removal of patients from the heart–lung bypass machine. The "heart–lung machine" is a large complex device that externally maintains circulation, oxygenation, and warmth of blood for the patient during the period of cardioplegia or when the heart has been stopped. The anesthesiologist, surgeon, and perfusionist work together to "go on" and "come off bypass." This is done within a very specific set of procedural guidelines, as the risk of injury (i.e., clotting, hemorrhage with exsanguination, air embolization) is very high during this part of the operation. It must be performed in a highly focused and coordinated manner between two members of different subteams who must come together to form an *ad hoc* subteam divided geographically based on degrees of sterility.

Methods for Enhanced Information Exchange during Cardiac Surgery: Implementing to Complement Existing Work

From a human factor perspective, the cardiac OR is a highly complex and tightly coupled environment during which failures are highly consequential for both the patient and the team. Successful operations in this specialty are dependent upon more than an individual technical skill. It is also heavily dependent upon communication across and within subteams, which can be very difficult to maintain due to cultural, environmental, as well as cognitive, technical, and physical workload. Within domains such as the military, briefings, in which teams meet to prepare for action, are regularly conducted to enhance communication within and across teams. Briefings remain infrequent in health care despite being touted as an essential part of surgical care and being widely endorsed (WHO 2009). Potential reasons for their underutilization are many, ranging from a myopic stance within each surgical discipline as to the uniqueness of their work, poor experiences in the past with "off-the-shelf" protocols that may not have been adequate (Catchpole and Wiegmann 2012), as well as logistical barriers such as scheduling and limited resources manifesting as individual resistance (Wiegmann et al. 2010).

From 2006 to 2008, a group of surgeons, nurses, anesthesiologists, perfusionists, and human factors and safety experts undertook a series of projects to design and evaluate team communication improvements, specifically that of preoperative briefing and a standard communication protocol designed especially for cardiac surgery. The full results of the study have been published elsewhere (Henrickson Parker et al. 2009; Wadhera et al. 2010); therefore, the following discussion is dedicated to understanding the practical considerations and method that were developed to achieve the results of the project. The critical aspects here are not the results but rather the path taken to achieve the results. The methods used here are not particular to cardiac surgery and may be applied to other health care domains. Because of the applied nature of this project, it does not strictly adhere to the tenants of a standard CWA. The interventions discussed in the following were very deliberately designed to account for the functional work of individuals within a complex cognitive system and focused on using a descriptive approach to identifying strategies of caregiver team performance and information sharing.

Design and Implementation of a Preoperative Briefing

The following discussion is focused on the method of designing of the briefing protocol, the implementation of the briefing, and the design of the measures. In line with the principles of CWA, the expertise of frontline staff was an integral part of the development of this intervention. In order to understand the influence of the system and the relationship of its components to communication and team performance, a preimplementation–postimplementation comparison was conducted, with process measures evaluating the effect of the briefing on communication as well as an iterative process to optimize the briefing and its implementation. The expertise of the staff, combined with the principles of human factors and safety science, resulted in both a meaningful intervention and meaningful measures.

Motivation

A team of human factor researchers were tasked not by the hospital or by the department of surgery but by individual surgeons and nurses to help facilitate better preparation for complex operations. They were motivated by desire to maximize safety and efficiency in their own OR. Our research team's goal was to first show value in one OR and then to build a more general approach throughout the cardiac ORs and ultimately across other surgical domains. To better understand the workflow of individual team members in the OR, prior to any data collection, familiarizing observations were conducted by the human factor researchers and medical students in the ORs. These observations, though not used for data collection, proved invaluable as they allowed the team to understand and appreciate the complexities of the work without the added pressure of data collection. In addition, it allowed the researchers time to appropriately define and scope the problem of intraoperative communication prior to crafting measurement tools.

Method

A structured survey was developed (Figure 8.3). The survey included openended questions and was administered by paper and pen at standing staff meetings. This was done with the support of the supervisory staff (e.g., nurse managers in cardiac surgery, the head of nursing for surgical services, the division chief for anesthesia, and the department chair for cardiac surgery) and was also included on the agenda at busy staff meetings. In these same meetings, the staff was led through a semistructured focus group designed to discuss their opinions on the design, logistical considerations, and usefulness of a preoperative briefing in cardiac surgery. Additional separate focus groups were conducted with nurses, surgical assistants, and surgical technicians, with perfusionists, anesthetic staff (certified registered nurse anesthetists and anesthesiologists), and surgeons. Each staff meeting and focus group was a single discipline, rather than all of the OR personnel in one room. This method was chosen deliberately to encourage staff from all disciplines to express their

---

Specialty: _____

Content of Briefing

1. What information/topics/issues should be discussed in a preoperative briefing from your perspective?
2. Is there information that others may know that you would like them to share with you in order to help the case run more smoothly? If so, who has the information, and what is it?
3. Is there information that you have that you don't normally share that you would like others to know?

Logistics

4. In your opinion, who should be present for a preoperative briefing?
5. When and where should it be done? (e.g., in the OR before the pause)
6. How long should it take?
7. Who should lead it?

Challenges

8. What barriers may exist that could prohibit a preoperative briefing?
9. Would you like to see some sort of preoperative briefing implemented in the OR?

FIGURE 8.3
Briefing opinion survey.

opinions either through writing or through discussion. The research team also chose this method to allow individuals who were unlikely to speak up among the strong hierarchy of their OR team an opportunity to share their expertise. Though this method was time consuming, ensuring that the major stakeholders (frontline staff) had an opportunity to discuss the content, structure, and logistics of the briefing was critical to the research team.

The survey asked specifically about how the staff felt about a briefing (attitudes); about their opinion on when, where, and how long the briefing should be (logistics); the content of the briefing; who should be involved in a briefing (participation); and potential barriers. Generally, the staff believed that briefings were a good idea and were supportive (attitudes) but that they had to be less than 5 min before each case and outside of the OR. There was significant variation in who should attend the briefings, with the nursing staff citing the largest number of attendees, and the surgeon and perfusion staff citing the fewest (surgeon, perfusionist, and anesthesiologist). The content of the briefing should cover both routine and unusual aspects of the procedure, the patient, and any equipment. Barriers were mainly focused around logistics ("everyone cannot attend," "too busy during preparation to do a briefing") and attitudes ("this isn't useful").

Developing a Prototype of a Preoperative Briefing Tool/Model

Based on the concerns and feedback of each of the different disciplines, a prototype preoperative briefing tool was developed. Initially, it was designed to focus on the procedure, the patient, and the equipment/resources, all of which would be discussed by the surgeon (Figure 8.4). The draft briefing was then shown to a small representative group of surgical staff from one OR.

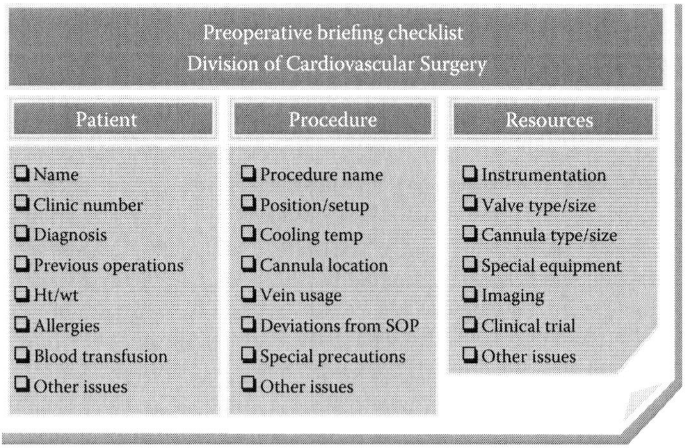

FIGURE 8.4
Iteration 1 of the preoperative briefing checklist.

The feedback from the initial design was that the briefing was too narrow and that the content was not useful as it was organized. In addition, the staff felt that the surgeon would be appropriate to lead the brief but that the design made it feel like the surgeon was the only one that could discuss any of the items on the list. They also felt that the briefing would only be complete if all the items on the list were "checked off" even if some items were not important to every case and that other items were missing that might be important for other cases. There was also concern among staff as to whether or not they would be audited for completion of the checklist.

For the research team, this was an incredibly useful phase of the research. Although the original intent of the briefing was to be a memory aid to facilitate conversation, this feedback highlighted that the design had implications that were significantly deeper (even subliminal) than originally intended. A number of changes were made based on this feedback, and multiple revisions were made based on additional feedback from this group and informal feedback from potential users.

To assess the impact of briefing, both qualitative and quantitative measurements were undertaken during prototype testing. Qualitative measures were used to iteratively improve the briefing protocol itself. The staff was continually asked to provide feedback, which was successively integrated. In addition, opinions about the briefing events were gathered anecdotally. After each design update, the briefing was again shown to the OR staff and iteratively assessed (Figure 8.5). Each prototype was presented to the teams in the morning before the first case and prior to patient arrival in the OR in which implementation was planned. For surgeons and anesthetic staff that were not able to be in the room prior to patient arrival, feedback was sought outside of the OR. All feedback was gathered via informal question and answer sessions and was iteratively integrated.

In the final version of the briefing, there were a few major changes from the original (Figure 8.4). First, there are more details on specific clinical issues, but they remain concise. This change was made because the OR staff felt that there was value in including these specific issues but wanted to keep the briefing short. The second major change is the organization of the briefing, from content-specific to individual-specific. The initial organization was very generic; however, clinicians did not find it useful in this format. Their feedback helped to iteratively move from a general checklist to a specific briefing with a structure consistent with the work as it was being done by a staff member. Based on research from high-reliability organizations, each member staffing a surgery case was asked to discuss particular aspects of the case specifically related to their area of expertise (e.g., the anesthesiologist assessment and concern about blood pressure control, the surgeon, and anatomic issues that may prolong the case, etc.). Each team member was expected to come to the briefing prepared, whereas in the initial design, the responsibility for conducting each section of the briefing was unspecified and therefore would likely have fallen to the surgeon or the surgical resident to perform.

OR Preoperative briefing trial version 5

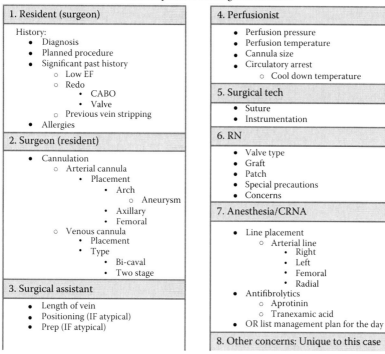

FIGURE 8.5
Iteration 5 of the preoperative briefing.

In that case, the briefing would not have been useful for the entire team as it would only have included input from one team member.

Finally, the item originally labeled "Other" at the end of the briefing was expanded to be its own section, labeled "other concerns unique to this case." The staff found this item particularly useful as it facilitated a "one-off" conversation about the case. The language was more specific than the "other" and therefore made the intended purpose of this section clearer. For example, in the testing period, during this part of the briefing, it was mentioned that the patient was schizophrenic and would need special care when woken from anesthesia. Though multiple individuals (anesthesia physician, circulating nurse, and surgeon) knew this information about the patient, sharing it with the room enabled the entire team to be prepared.

Prior to the final briefing being integrated into practice, ethnographic observations were conducted by a medical student in the implementation OR. To determine what measures would be gathered, a number of communication measures from the literature were included as well as measures that were relevant to the staff. The frequency and type of communication issues (procedural knowledge, equipment preparation, and miscommunication), the frequency of trips by the circulating nurse trips to the supply room, as well as the amount of wasted (or opened but unused) disposable equipment were all gathered. These measures showed process and efficiency improvement and perceived frustration levels by staff. For example, in discussions with the implementation OR staff, multiple individual members mentioned the nurse being out of the room when she or he was needed ("The nurse is always out of the room when I need something!"). Therefore, we hypothesized that a functional preoperative briefing would improve awareness of possible needs for the case and thus decrease unexpected trips to the supply area.

After the prebriefing observation period was conducted, the briefing tool was finalized and implemented. It was a single-sided sheet of laminated paper for ease of cleaning and reuse. The briefing was conducted in the hallway outside the OR at an appointed time (7:30 a.m.) for only the first case of the day. The same observer conducted postimplementation observations using the same measures. In the postimplementation period, an additional observer came to the briefing to help with any questions or concerns and to assess the briefing itself (e.g., which staff attended, when did it start, how long did it take).

This design process represents the benefit of CWA in action: designers of systems can use this method to move beyond imagining or assuming how work is or should be conducted and designed in response to how individuals actually accomplish their work goals, which is often quite different than assumed. Designing the intervention based on clinicians' iterative feedback allowed the development of a communication tool that enhanced and supported the "work as done." The application of cognitive engineering methods also enabled this intervention to improve teamwork and safety while being useful, effective, and appropriate for the clinical environment, the ORs, which are complex, high-stress, time-pressured, and high-consequence environments.

Dissemination and Sustainment

All of the data from this study related to the utilization of the tool were based on briefings conducted at the beginning of the surgical work day, before the first case of the day, and in one OR with one OR team. Interestingly, once briefings were integrated into the practice for the first case, the personnel comprising the subteams assigned to the experimental OR requested briefings with this surgeon for the next cases that they worked together. What started as a briefing for a single case turned into a briefing for multiple cases in multiple ORs utilized by these workers. This development was organic and, though not included in the research project, is believed to be indicative of the value of the briefing itself and of the method for developing an effective briefing that meets the workers' needs. Also of interest was the decision by the facility not to integrate the preoperative briefing into the standard written or electronic medical record. This decision was widely supported by the clinical staff and by the research team, because it ensured that the structure could remain fluid and did not become a tool for auditing purposes. By placing it in the medical record, it takes on a different purpose than it was originally intended. Integration into the medical record makes it more focused on documentation accuracy, whereas the purpose of the tool was to facilitate team communication and coordination. The purpose of the briefing designed by and for the staff was to facilitate conversation and information sharing, different from a "time out" that is a mandated information delivery, by accrediting agencies, to occur prior to the first incision. The preoperative briefing toll was designed with an emphasis upon facilitating a dialog across and between all team members to create a shared mental model of the operation and to give each team member an opportunity to calibrate to common ground as quickly and thoroughly as possible in a nonthreatening environment.

Though the preoperative briefing was highly successful in one OR with one surgeon, dissemination was challenging. Some surgeons readily adopted the briefing, while others did not, unconvinced of the benefits despite data to suggest otherwise. The nature of surgical work results in the surgeon having a high level of autonomy within their ORs, and as such, this type of resistance was anticipated. Challenges remain to acceptance of any enhancement to clinical work; however, this study demonstrated the power of understanding the work environment, the barriers, and the limitations, and the meaningful measures were imperative for success with a single team. It also shows how closely one can match the enhancement to the actual work. In this case, it fueled an extemporaneous unplanned dissemination by those who are members of the test cardiac surgical team to ask for the briefing because they found it usable and, as such, worthy of adoption and dissemination.

Design and Implementation of a Surgeon–Perfusionist Communication Protocol

As part of the larger work (Wadhera et al. 2010), the team was asked to use similar techniques to design a communication tool to enhance surgeon and perfusionist communication, regarded as a critical failure point within cardiac surgery. The expertise of frontline staff was used extensively, and an understanding of how the work system influences this communication was key for the design.

Motivation

Within a cardiac operation, one of the most critical communication axes is between the surgeon and the perfusionist who runs the heart–lung bypass machine. Often referred to as the "pump," this machine does the work of the heart and lungs during the cardiac operation by acting as external heart and lungs. The machine pulls the patient's blood from the body through the pump (bypassing the patient's heart and lungs), oxygenating and warming the blood in the process, and then returning the blood back into the body (Figure 8.6). Connected directly into the large vessels of the body, this is a large piece of complex technology that must be wheeled into and out of the OR (Figure 8.2).

The heart–lung bypass machine is run by a perfusionist who has received special education and training specific to the use of this device during cardiac surgery. The physical location of the pump, and its operator, is variable and dependent upon the size of the OR. It may sit off to the side, in front, below, or even behind the surgeon. Despite being situated in the nonsterile zone along with its operator, the tubing extending from the pump must cross into the sterile field and be attached to the patient.

The work of a perfusionist is one of long periods of vigilance punctuated with moments of intense cognitive and physical activity. There are very specific pieces of information that the surgeon and the perfusionist need to

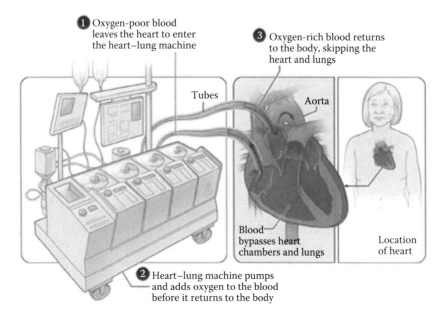

FIGURE 8.6
Heart–lung bypass machine.

Surgeon	Perfusionist
Is the vent on or off?	
	Vent is on 500.
I thought the vent should be off.	
	I turn it on after cardioplegia.
You shouldn't have done that because the LV clamp was off.	
	Sorry about that.
Next time tell me what you're doing.	

FIGURE 8.7
Typical communication exchange.

share in order to put the patient onto cardiopulmonary bypass, or stop the patient's heart, and to take the patient off and restart the heart. At the time of this study, commands were unstructured with each surgeon and each perfusionist presenting information in a nonstandard way (Figure 8.7). Both perfusionists and surgeons felt that there were significant risks associated with the nonstandard communication and wanted to design a communication protocol to aid standardization.

A few things about the exchange in Figure 8.7 stand out. First, there is no obvious system indicator of the current status of the system ("Is the vent on or off?" and "I thought…" and "You shouldn't have done that because…"). Second, the phrase "I turned it on after cardioplegia" is indicative of the idiosyncratic nature of work with this machine. In these ORs, the work was entirely unstandardized. Third, there is an anticipation that this same error will occur again ("Next time…"). This brief exchange illustrates that there are significant opportunities for improving communication during this high-risk procedure.

Method

First, it is important to identify what verbal interactions occur between the surgeon and the perfusionist using a "communication map," which was created by the human factor experts and perfusionist for each different type of cardiac surgery utilizing the pump. The communication map was designed to provide a "one-look" visualization of each of the critical bypass stages and the required communication at each stage. The maps created for each type of cardiac surgery (e.g., aortic valve replacement, coronary artery bypass grafting [CABG]) included all the operative stages of each different type of procedure (e.g., administration of anesthesia, cardiopulmonary bypass [CPB], cannulation of large vessels of the body to attach to the pump, etc.). Communication exchanges between the surgeon and the perfusionist at each critical stage were written independently by five experienced perfusionists, combined by a human factor researcher, and then reviewed by three cardiac surgeons. Then a comparison across each surgery was done to identify specific instances of critical communication that were consistent no matter the type of cardiac procedures being performed (Figure 8.8). Direct observations confirmed the existence of all the stages and the information exchanged at each stage.

After all the communication maps were created and revised, a focus group of both surgeons and perfusionists was convened to determine exactly what particular pieces of information are critical to both parties. During this meeting, discussions of what critical information was needed, how that information should be exchanged, and the level of detail desired by each party were determined. For example, one exchange between a surgeon and a perfusionist showed very different mental models, priorities, and expectations of how to get work done. The perfusionist said, "The surgeon takes the (aortic) root vent (catheter) out, which isn't a critical step for him, but without the root vent, the perfusion reservoir (on the machine) can run dry (of blood)." Failure to communicate with the perfusionist that the catheter located in the aortic root has been removed can have an effect up the reservoirs of blood and fluid that are still connected to and circulating through the patient's other major blood vessels. Based on these findings, a protocol was developed (Wadhera et al. 2010).

Using the same interactive process as with the briefing tool, the final protocol surgeon–perfusionist communication tool included eight generic critical stages that require a direct verbal exchange between the surgeon and the perfusionist:

1. Establishment of ACT
2. Circuit check

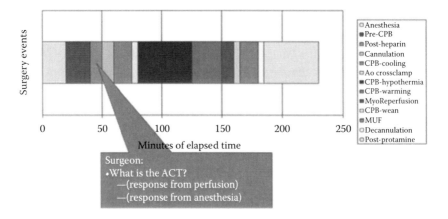

FIGURE 8.8
Cardiac surgery timeline events for CABG (ACT is activated clotting time, an indicator of coagulation of the blood after heparin administration that tells the operating team whether or not the patient is adequately anticoagulated and thus ready for the operation to begin).

3. Initiation of CPB

4. Cross-clamp on

5. Administration of cardioplegia

6. Vent on and off

7. Cross-clamp off

8. Termination of bypass

For example, in practice, step 1 of the protocol requires that the ACT level be given rather than a statement that "ACT is adequate" in support of cross-monitoring. Further, a confirmatory callback of all exchanges was instituted as required of both parties. A specific script was not created for every type of cardiac surgery, but the format of the exchanges was mandated for use whenever the heart–lung cardiac bypass procedure is being utilized.

Before implementation of the protocol, a convenience sample of 18 operations using the pump was observed, and every communication exchange between the surgeon and the perfusionist was recorded as close to word-for-word as possible. Each communication exchange was then classified as either no issue or as one of the following types of communication breakdowns: (1) miscues, (2) no call-back (either surgeon or perfusionist), (3) repeated communication exchanges, (4) occurrence of a nonverbalized critical action, and (5) ambiguous or unstructured communication exchanges (Table 8.1). These categorizations were determined before observations on the basis of our previous work in the cardiac surgical OR. Categorization of each communication event was validated by independent categorization by two raters, and an agreement was determined. Observed communication exchange data were also used to determine the frequency at which no communication or poor communication occurred during each critical stage. After the

TABLE 8.1

Definitions of Communication Breakdowns

Type of Breakdown	Definition
No call-back	Exchanges between the surgeon and the perfusionist in which a confirmatory call-back of action or instruction did not occur.
Repeated communication	A communication exchange that must be repeated.
Nonverbalized critical action	Occurrence of actions that were not communicated between the surgeon and the perfusionist.
Ambiguous or unstructured communication exchange	Communication exchanges that do not include specific instructions (e.g., cardioplegial to be given antegrade or retrograde).
Miscues	Surgeon or perfusionist not clearly hearing communication exchanges or requests from on another either because of volume or because of confusion.

implementation of the protocol, the same measures were taken in the ORs with a convenience sample of 16 cases.

The eight critical categories of the surgeon–perfusionist communication tool were printed on a small laminated poster and taped to the wall in the OR in the field of view of the surgeon, and were also given to the perfusionist to put at the pump station. Many perfusionists taped the protocol to the bypass machine. During early implementation, the surgeons would frequently look up at the protocol, but after using it for a couple of operations, they no longer looked at it but were still verbalizing all the eight critical periods.

It is important to note that our communication tool ended up not needing repeated iterative revisions as the briefing tool. We believe that this is because of the high risk required by the stages of the pump use, which are prescriptive for correct usage of the device, and as such there was less room for interpretation. Additionally, the research-design team had become more facile at integrating user feedback earlier in development for an expedited final design. As seen with the briefing tool, the use of a cognitive engineering, work-centered approach yielded a protocol that was easily interpreted and integrated by the frontline staff. By ensuring a user-based and work-centered process, stress and frustration were reduced all around. It also resulted in a better quality, more compatible communication tool being implemented rather than the pilot-air traffic control tools being considered by the organization prior to the study.

Dissemination and Sustainment

The communication protocol was integrated into six different ORs, and there were significant improvements in communication across all the ORs (Wadhera et al. 2010) and across all communication breakdown categories except for miscues. Using the method of integrating frontline feedback and engaging all of the stakeholders throughout the development of the protocol enhanced its utility. The protocol was utilized in the ORs even after the study period ended. It is not known whether the protocol is still in use four years later, as all members of the study team are new at different facilities.

Conclusion

The introduction of any changes to the surgical health care work setting, no matter how well-meaning, will have obvious and latent consequences. These are often identified retrospectively and difficult to predict without a work analysis that includes physical staging of the work and an attention to the cognitive work occurring. This discussion has deliberately been focused on a single subsystem—cardiac surgery—as a model for the need of similar prospective activity throughout health care prior to implementation of proposed "enhancements" and "improvements." The evidence from the work reviewed here related to preoperative briefings and surgeon–perfusionist communication protocols demonstrates the importance of iterative collaboration among human factor experts, cognitive engineers, and clinical workers (Perry 2004). Adoption of new work tools, no matter the domain, is based on the tool assisting in the realized needs and risks workers are coping with. Cultural and behavioral changes within health care will continue to be difficult to implement and sustain without an appreciation of the complexity of the work and how workers function in a work environment. Adaptive partnerships among human factor experts, clinical workers, and administrators offer the best opportunity for designing cognitive and communication tools that match the "work as it occurs" and ultimately adopting them by frontline workers.

References

American College of Surgeons. (2004). American College of Surgeons Statement on Principles. *Statement on Principles.* Retrieved from http://www.facs.org/fellows_info/statements/stonprin.html#pre.

Behara, R., Chinander, K., Barreto, C., Perry, S. J., and Wears, R. L. (2005). Analyzing human performance in knowledge-intensive services: A study in emergency care. *Proceedings from the 36th Annual Meeting of the Decision Sciences Institute, San Francisco.*

Carayon, P., Schoofs Hundt, A., Karsh, B. T., Gurses, A. P., Alvarado, C. J., Smith, M., and Flatley Brennan, P. (2006). Work system design for patient safety: The SEIPS model. *Quality and Safety in Health Care, 15 Suppl 1,* i50–i58.

Catchpole, K., and Wiegmann, D. A. (2012). Understanding safety and performance in the cardiac operating room: From "sharp end" to "blunt end." *BMJ Quality and Safety, 21*(10), 807–809.

de Leval, M. R., Carthey, J., Wright, D. J., Farewell, V. T., and Reason, J. T. (2000). Human factors and cardiac surgery: A multicenter study. *Surgery, 119*(4), 661–672.

Edmondson, A. (2003). Speaking up in the operating room: How team leaders promote learning in interdisciplinary action teams. *Journal of Management Studies, 40*(6), 1419–1452.

Grill, C. (2012). State of the States: Defining surgery. *Bulletin of the American College of Surgeons, 97*(5), 27–29.

Henrickson Parker, S., Wadhera, R., Elbardissi, A. W., Wiegmann, D. A., and Sundt III, T. M. (2009). Development and pilot evaluation of a preoperative briefing protocol for cardiovascular surgery. *Journal of the American College of Surgeons, 208*(6), 1115–1123.

Henrickson Parker, S., Yule, S., Flin, R., and McKinley, A. (2012). Surgeons' leadership in the operating room: An observational study. *American Journal of Surgery, 204*(3), 347–354.

Lingard, L., Espin, S., Whyte, S., Regehr, G., Baker, G. R., Reznick, R., Bohren, J., Orser, B., Doran., D., and Grober, E. (2004). Communication failures in the operating room: An observational classification of recurrent types and effects. *Quality and Safety in Health Care*, 13(5), 330–334.

Mathieu, J. E., Marks, M. A., and Zaccaro, S. J. (2001). Multi-team systems. In Anderson, N., Ones, D., Sinangil, H. K., and Viswesvaran, C. (Eds.), *International Handbook of Work and Organizational Psychology*, pp. 289–313. London: Sage.

Perry, S. J. (2004). An overlooked alliance: Using human factors engineering to reduce patient harm. *Joint Commission Journal on Quality and Safety*, 30(8), 455–459.

Pilcher, L. (1917). Definition of major and minor surgery. *Annals of Surgery*, 65(6), 799.

Reason, J. (1995). A systems approach to organizational error. *Ergonomics*, 38(8), 1708–1721.

Reason, J. T. (1997). *Managing the Risks of Organizational Accidents*. Aldershot, UK: Ashate Publishing Co.

Shortell, S. M., and Singer, S. J. (2008). Improving patient safety by taking systems seriously. *JAMA*, 299(4), 445–447.

Undre, S., Healey, A. N., Darzi, A., and Vincent, C. A. (2006). Observational assessment of surgical teamwork: A feasibility study. *World Journal of Surgery*, 30, 1774–1783.

Vicente, K. J. (1999). *Cognitive Work Analysis: Toward Safe, Productive, and Healthy Computer-based Work*. Mahwah, NJ: Lawrence Erlbaum Assoc Inc.

Vincent, C. (2003). Understanding and responding to adverse events. *New England Journal of Medicine*, 348(11), 1051–1056.

Vincent, C., Moorthy, K., Sarker, S., Chang, A., and Darzi, A. (2004). Systems approaches to surgical quality and safety from concept to measurement. *Annals of Surgery*, 239(4), 475–482.

Wadhera, R., Henrickson Parker, S., Burkhart, H., Greason, K., Neal, J., Levenick, K., Wiegmann, D., and Sundt III, T. M. (2010). Is the "sterile cockpit" concept applicable to cardiovascular surgery critical intervals or critical events? The impact of protocol-driven communication during cardiopulmonary bypass. *Journal of Thoracic and Cardiovascular Surgery*, 139(2), 312–319.

Wears, R. L. (2003). A different approach to safety in emergency medicine. *Annals of Emergency Medicine*, 42(3), 334–336.

Wears, R. L., Perry, S. J., and Sutcliffe, K. M. (2005). The medicalization of patient safety. *Journal of Patient Safety*, 1(1), 2–6.

World Health Organization (WHO), Patient Safety Group (2009). WHO Guidelines for Safe Surgery 2009: Safe Surgery Saves Lives. Geneva, Switzerland: WHO Press. Available at http://whqlibdoc.who.int/publications/2009/9789241598552_eng.pdf

Wiegmann, D. A., Eggman, A. A., Elbardissi, A. W., Parker, S. H., and Sundt III, T. M. (2010). Improving cardiac surgical care: A work systems approach. *Applied Ergonomics*, 41(5), 701–712.

9

Engagement and Macroergonomics: Using Cognitive Engineering to Improve Patient Safety

Yan Xiao and C. Adam Probst

CONTENTS

Introduction

Safety improvement in health care organizations can benefit greatly from cognitive engineering concepts and methods, especially in areas such as engaging key stakeholders in using multiple leveraging points such as policy, education, and cultural changes. In contrast with more traditional domains in which cognitive engineering is applied, such as aviation and device design, health care organizations are complex sociotechnical systems with constraints that limit changes to devices, to interfaces, to physical plant layout, or in selecting employees. Cognitive engineering has the potential to create solutions under these constraints. Four cases in one health care organization are utilized to illustrate strategies and provide a demonstrative framework regarding the use of cognitive engineering methods and concepts.

Background

Safety and efficiency in health care systems are emergent properties that stem from factors found in large sociotechnical systems. Cognitive engineering brings useful perspectives, tools, and methods that can affect those factors at multiple levels of a healthcare organization, ranging from the design of administrative dashboards for executives and managers to education materials for patients and families. The macroergonomic perspective is very relevant to applications of cognitive engineering, as problems are rarely solvable by working on a single component such as redesign of a user interface. In this chapter, four case studies are used to illustrate how cognitive engineering methods and concepts are used to address challenges and solve specific problems faced by a large health care delivery system.

Baylor Scott & White Health is a large, integrated health care delivery organization with multiple hospitals and primary care networks. A number of improvement projects have benefited directly from concepts and methods often employed by cognitive engineering. One key strategy that has been used is to consider multiple points of leverage within the sociotechnical system such as policy, education, and cultural changes. In contrast with more traditional application domains such as aviation and device designs, health care organizations tend to have more stakeholders and are more limited in making changes to devices, interfaces, physical plant layout, or employee selection. In order to have an impact, cognitive engineering must be deployed under these constraints and bring viable, sustainable solutions to the table. Baylor Scott & White Health has two full-time, doctoral trained human factors specialists with training and experience in cognitive engineering. They are called upon to address important, usually challenging problems facing the organization. Cognitive engineering methods and concepts—in particular, a commitment to understanding work in context, as experienced by frontline care providers, and integrating multiple levels of the sociotechnical system in order to develop and deploy solutions—have proven to be invaluable in identifying improvement opportunities and in developing solutions.

Case 1. Nursing Workload

An evaluation of intensive care unit (ICU) patient safety studies revealed that nursing workload is a direct contributor to safety and the quality of care (Carayon and Gürses 2005). The direct link among nurse workload, patient safety, and the nurse's well-being has been well documented (Aiken et al. 2002; Kazanjian et al. 2005).

Historically, standard workload measures do not promise gains in nursing efficiency nor are they capable of adequately capturing the true complexity of nursing workload (Brady et al. 2007; Morris et al. 2007; Weydt 2009), especially the issues attributable to the work environment. These work environment variables have been the least studied aspect of nursing workload (Neill 2011), and minimal attention has been focused on the actual mental workload requirements of nurses to meet productivity, quality of care, and patient safety metrics (Aiken et al. 2002; McGillis-Hall 2003; Carayon and Gürses 2005). Standard measures of nursing workload have traditionally relied on skill competency, task performance, and time required to complete tasks while omitting the contribution and insight that human factors and cognitive engineering can provide (Neill 2011). However, in order to maximize the impact of improvement strategies, an intimate understanding of the entire work system is required, informed by human factors engineering, that takes into account all aspects of the sociotechnical system, including cognitive workload (Reason 2000; Page 2004; Karsh et al. 2006; Carayon 2007).

Understanding Workload

The most common method to evaluate nursing workload, staffing (such as measured by patient/nurse ratios), is not truly representative of the actual workload experienced by nurses (Adomat and Hicks 2003; Pearson et al. 2006). A more accurate definition of workload is the ratio of the demands placed upon nurses (i.e., the task load) to the available resources the nurses possess to accomplish those tasks (Xie and Salvendy 2000). One such resource is the cognitive ability demanded of nurses to adequately process vast amounts of information and to remember key tasks (so-called prospective memory).

A human factors framework provides a more insightful illustration of the actual workload that nurses experience, as imposed by the work system and

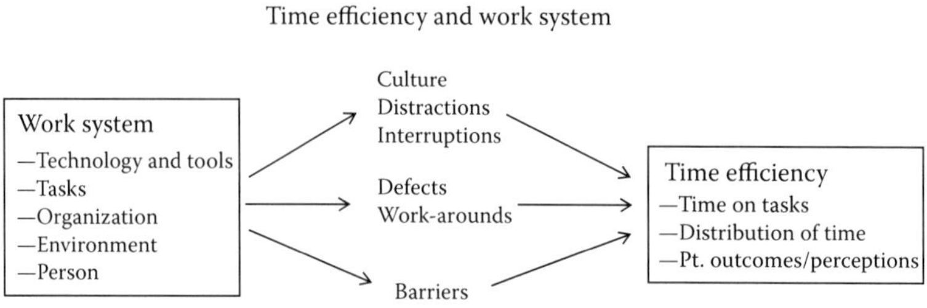

Time efficiency and work system

FIGURE 9.1
Work system framework used to educate project team members on a systems engineering approach to efficiency. The figure was utilized to engage leadership and team members to think more broadly about how people, technology, tools, and the environment together comprise the work system and interact with current culture, work-arounds, and barriers to impact nurse efficiency.

often exacerbated by suboptimal work system design choices. Cognitive engineering methods are very useful to reveal opportunities for improvement in the work system. One example task is medication administration, in which the workload involved is dependent on tools, physical layout, user interfaces, other concurrent tasks, interruptions, and skills in using the tools provided. In other words, a human factors approach to workload encourages the examination of factors that are not usually reflected by the more traditional nursing workload indicators. Workload should be examined by aspects of the entire work system such as the interruptions incurred by nurses, divided attention demanded, and the time pressure to complete tasks (Holden et al. 2011).

In this case, the human factors team was called upon to address a sharp decline in nursing responses of job satisfaction indicators in one of Baylor Scott & White Health's facilities. Despite no significant change in patient/nurse ratios, nurses rated their job enjoyment, staffing and resource adequacy, and perceptions of the environment noticeably lower than in the previous years. Based on the impact of previous participation of human factors specialists, the facility leadership requested that a human factors evaluation of current nursing workload be completed in order to identify and resolve nursing pain points. We created a work system framework for the project modeled after SEIPS (Carayon et al. 2006), as shown in Figure 9.1.

Methodological Considerations

We decided to leverage the fact that rapid-cycle improvement is a widely used methodology within the organization. Rapid-cycle improvement encourages successive adjustment of improvement interventions based on ongoing evaluation. Cognitive engineering methods are well suited for different types of activities during rapid-cycle improvement initiatives. The facility's chief nursing officer enlisted two units as targeted pilot care areas. One was a medical/surgical unit staffed with experienced tenured nurses and the other an ICU. We partnered with the targeted units' nursing managers to identify their assessment of the staff's demanding tasks and then confirmed and revised their suggestions with informal staff interviews. By this process, we determined that the first 4 h of a nurse's 12-h shift contained a multitude of high cognitively demanding tasks: shift change and bedside report, initial head-to-toe assessment of each assigned patient, the first round of medication administration (which contained the bulk of scheduled medications to be administered to patients on that shift), and the initial documentation requirements via the electronic health record (EHR). Focusing on the first 4 h of a typical nursing shift limited the scope of the project and made it feasible to be carried out with limited resources and time. Figure 9.2 illustrates how the targeted first 4 h related to a 12-h nursing shift. Based on the work system framework (Figure 9.1), we expected to uncover tasks performed by nurses related to shift change, medication administration, documentation, and supply management (e.g., gloves, syringes, alcohol swabs, etc) as displayed in Figure 9.2.

Cognitive engineering concepts and approaches provide a framework to identify, understand, and more accurately capture a more representative measure of nursing workload in contrast to the more standard nurse/patient ratios. In addition to time and motion methodology, which was necessary to map and quantify nursing workflow, we utilized semistructured interviews. These interviews were designed to illicit instances where the perceived workload was high and identify contributing factors that prevented working efficiently. Additionally, we asked nurses to describe the tasks they believed provided little to no value to patient care. Furthermore, we captured and evaluated artifacts created by nurses to assist in their completion of tasks. For example, ICU unit nurses utilized a custom "Jot Sheet," which allowed them to track a patient's progress, vital signs, and critical lab results and served

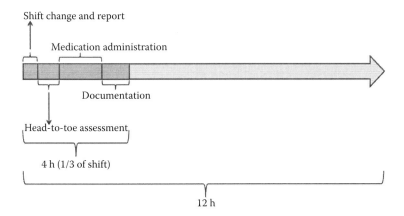

FIGURE 9.2
Timeline of the first 4 h in a standard 12-h nursing shift. Staff interviews revealed that nurses spend around the first 4 h (i.e., 1/3) of their shift completing multiple challenging tasks in rapid succession. All of these tasks are individually demanding and require extensive cognitive resources with significant time requirements that contribute noticeably to the nursing workload.

as a reminder for upcoming patient care events such as imaging studies or procedures. We also took notice of the teamwork structure that nurses had implemented that provided them extra support for high-acuity patient care tasks when needed and allowed the opportunity for breaks when possible.

In addition to semistructured interviews, we created a survey for nurses to report interruptions and distractions they endured throughout the medication administration and EHR documentation process (represented in Figure 9.2). The survey served as a baseline subjective assessment of how nurses viewed the impact of interruptions to their workflow and a rating of the primary causes of those interruptions (e.g., ancillary departments, other nursing staff, physicians). We used photographs for illustrating and communicating defects in the work system, as well as any improvement at remedying such defects. Although the term "cognitive engineering" was not used formally in project discussions due to the multidisciplinary team consisting of individuals not formally trained in human factors, cognitive engineering methods and concepts were employed and provided a more nuanced understanding of nursing workload. As a result, the project team gained more insight into barriers and tasks that result in increased demands on nurses and other frontline staff than what would have typically been possible.

Macroergonomics: Importance of Engagement, Teamwork, and Communication

It is important to recognize that frontline workers and management are the owners of nursing workload rather than the corporate or team members who contribute both positively and negatively to the workspace in which nurses must complete patient care. To truly make an impact in health care organizations, cognitive engineering applications must be based on maximum engagement at both the organization's executive leadership level and at the frontline staff level. In this case, the human factors team prepared short, "bite-size" orientation materials about human factors (e.g., the work system framework in Figure 9.1) with executives and frontline managers. Photographs and case examples were used to show cognitive engineering perspectives. Observation and improvement activities were designed and carried out jointly with nurses. Identified improvement opportunities were communicated and shared with those who owned the relevant processes (e.g., the pharmacy director of the facility on issues with medication availability and communication of missing doses). Additionally, in order to develop a multidisciplinary team that would be able to adequately handle a project of this complexity, we created two teams: a project team and a project-steering team. The project team consisted of the facility's Chief Nursing Officer (CNO), the vice presidents of medical/surgical nursing and critical care nursing, the managers and one frontline nursing representative from each unit, nursing EHR support staff (who were nurses themselves), human factors specialists, and the facility's patient safety officer. The steering team was composed of the system and facility CNOs, the associate chief quality officer, the vice president of patient safety for the region, the director of multidisciplinary informatics, and human factors specialists. Throughout the study, the project team met biweekly to discuss new insights and identify potential solutions, whereas the steering team met monthly to ensure the study's continued momentum, remove barriers for intervention implementation, and martial resources for the project team.

Contributing to Solutions Using Cognitive Engineering

Due to economic and governmental shifts in the field of health care, Lean and Six Sigma methodology are often employed in an attempt to reduce waste and provide a means to offer patient care that uses resources more efficiently. While those approaches have increasingly been used in health care systems, they are not "one-size-fits-all" solutions in health care environments. Nursing work has been impacted by the use of information technology in health care, where vast amounts of data are at the fingertips of health care professionals; providers are being asked not only to find and assimilate these data but also to accomplish increasingly complex tasks on a higher-acuity patient population. In this case, the utilization of cognitive engineering approaches in addition to more common approaches used to study hospital workflow such as Lean or Six Sigma allowed frontline users to identify bottlenecks that resulted in increased workload. For example, we looked specifically at how cognitive work was supported in the EHR-enabled work environment. One example was the lack of use of the EHR during shift change. The EHR did not provide nurses an adequate patient summary and was found to be left powered off during shift report (Figure 9.3a). Another example was a defect in the work system in adapting to using the EHR. We uncovered that the placement of the computer installed for inpatient

 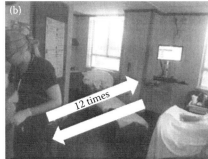

FIGURE 9.3
Pictures were used in discussions about barriers to nursing efficiency among team members. (a) Photograph shows that the EHR system was not used due to lack of an adequate patient overview to meet the demand of assimilating vast amounts of patient information at shift change. (b) Photograph shows a nurse had to walk back and forth 12 times between the medication preparation area and the computer workstation in order to interact with the bar code medication administration system.

rooms forced nurses to walk back and forth multiple times while using the bar code scanner during medication administration. We observed that nurses were often required to crush pills, draw up syringes, and mix dietary meal replacement solutions while interacting with the bar code medication administration program in the EHR (i.e., alerts and data entry requirements). In one observation, a nurse was observed to walk back and forth between preparing medications and the EHR 12 times (Figure 9.3b).

Discussion

Workload is a problematic issue in today's health care environment as nurses are continually asked to perform more increasingly complex tasks on sicker and sicker patients than they can adequately and safely accomplish, with tools that do not always follow a user-centric design all while having fewer and fewer resources to complete their work. The cognitive engineering approaches illustrated above, paired with the strategy of maximizing leadership and staff engagement and project ownership provided a framework that allowed the workflow and workload of nurses on two units to be well understood, mapped, and quantified in a meaningful context, which led to the implementation of targeted intervention strategies that improved the efficiency and reduced the workload of frontline nurses.

Case 2. Tubing Misconnections in Cardiovascular Surgery

Cardiovascular (CV) surgery taxes the operative team's cognitive resources during critical transitions throughout a surgery and requires perfect team communication (Wadhera et al. 2010). Human factors and systems engineering approaches have been utilized to identify hazards during CV surgery (e.g., Gurses et al. 2012). After a near miss during cardiopulmonary bypass (CPB), one facility in our organization requested an internal audit by human factors specialists in order to identify opportunities for safety enhancements to both prevent the near miss from happening again and proactively avoid potential future safety threats. Previous human factors analysis has highlighted usability challenges of CPB machines (Wiegmann et al. 2009).

Methodological Considerations

After engaging with the core staff of CV surgery, we began with a cognitive task analysis to identify the needs and strategies of surgeons, anesthesiologists, and perfusionists (the highly trained staff that operate the bypass machine). By walking through scenarios with each clinical group, we were able to identify and analyze the individual tasks and work activities needed to successfully hook up, run, and safely disengage a CPB machine from a patient. Once the cognitive tasks and user expectations were clearly understood, direct observation of multiple CV surgical cases allowed us to identify potential safety gaps that can arise during these complex cases. During these observations, we queried staff before, during (within approved times when the workload is low, and no risk to the patient was present), and after each case about variations in observed approaches. At the time of the study, a new group of CV surgeons had just begun practicing at the facility in which the observations occurred. These new surgeons' methods conflicted with those of the established CV surgeons that had been in practice for some time. The variations increased the burden on staff as well as introduced communication barriers resulting from differences in terminology. Interviews with staff identified each user's needs and expectations and their perceived opportunities for risk. The discussions and observations increased engagement with the clinicians involved and laid a foundation for an iterative approach to solution identification and implementation.

Contributing to Solutions Using Cognitive Engineering

Through employing cognitive engineering approaches, the team was able to identify several gaps in safety and potentials for error. For example, we mapped out the complex task sequence required to hook up the CPB machine to the patient prior to starting bypass to understand the cognitive demands placed on both the surgeon and the perfusionist. Additionally, through focused observation, we were able to uncover potential mechanical, procedural, and workspace contributors to errors. One particular hazard discovered involved the risk of CPB tubing misconnections. CPB machines include arterial and venous lines that must be connected to the appropriate anatomical structures of the patient's heart in order to correctly cycle and oxygenate the patient's blood as it passes through the machine. However, the potential for confusion existed as both the arterial and venous lines employed for certain cases were identical in appearance and diameter (Figure 9.4a). Clinicians placed a few pieces of colored tape to identify the appropriate function for each line: using red tape for arterial lines and blue/green tape for venous lines, placed on the ends that connect to the patient (Figure 9.4b).

Tubing misconnections, in general, are a persistent common issue (Commission 2006). The most commonly recommended solution for misconnection issues is a complete redesign of the associated tubes and connectors (Simmons et al. 2011). However, it is not always feasible for a health system to realistically purchase entirely new sets of equipment, especially in cases involving particularly expensive equipment such as CPB machines. We decided to focus on labeling, which was found to be a proactive deterrence to misconnections (Kimehi-Woods and Shultz 2006).

The perfusion service supporting the CV surgery teams was able to engage the tubing manufacturer and arrange for prelabeled arterial and venous line tubing to ensure consistency (Figure 9.5). The manufacturer was able

FIGURE 9.4
Tubing used in a cardiac pulmonary bypass procedure. Photographs are a very effective method to communicate patient safety risks identified using cognitive engineering approaches. Mix-up of tubing for arterial and venous lines is a serious risk, but both types of tubing were located next to each other with markings that were difficult to distinguish. (a) Photograph shows no markings visible for the arterial line with venous line markings that are difficult to see. (b) Photograph shows that the arterial line markings were near the floor and only visible from a specific angle.

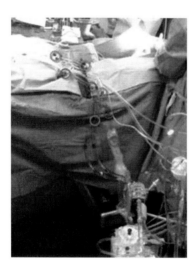

FIGURE 9.5
New tubing with revised labeling used during a cardiac pulmonary bypass procedure. Color-coded labels placed every 6 in. by the vendor helped prevent labels being accidentally removed when the tubing was connected to the patient. Before the revision, the labels were located only at the end of tubing and may be cut off when connected to the patient.

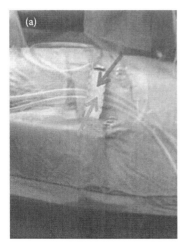
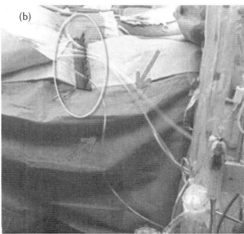

FIGURE 9.6
New manifold was purchased to physically separate the arterial and venous lines. (a) Photograph shows the old manifold with tubings located next to each other, making it difficult to distinguish the arterial line from the venous line. (b) Photograph shows the new manifold physically separating the arterial and venous lines and helped staff distinguish one line from the other. Pictures of interventions successfully implemented were used to communicate important improvements to patient safety and helped other facilities understand how their own operating rooms might accomplish similar interventions.

to provide this at no additional cost to the facility. Further, the team recommended that a new manifold be purchased and implemented to further separate and effectively distinguish the lines as they passed into the sterile field (before–after comparison in Figure 9.6a,b).

To address confusion in terminology, a comprehensive standardized "Cardiovascular Operating Room (CVOR) Instrument Book" was created, which contained a picture of every type of instrument used throughout the CVORs for every surgeon with the agreed upon, standardized name in addition to common synonyms. This book was then taught to each CVOR staff member. The standardization helped reduce unnecessary contributors to cognitive workload.

Discussion

Cognitive engineering tools and methodologies, such as cognitive walk-throughs, staff interviews, and field observations, facilitated the discovery of previously unrecognized opportunities for error in CPB machine use. Moreover, employing these tools allowed a framework for the multidisciplinary team to break down barriers in communication, which might not otherwise have been achievable. The solutions implemented to address identified safety gaps were developed and evaluated for efficacy based upon the extensive knowledge of human cognitive capabilities and limitations provided by the human factors team members.

Case 3. Efficiency of Documentation in EHR Systems

Documentation is a critically important activity for nurses and other professionals in health care. As a living document, a patient's chart supports individual cognition and team communication. For example, a patient's allergic responses to medications may be communicated to another team member via documentation in his or her chart. Documentation also plays an important role in the medicolegal and regulatory compliance realm. Increasingly, documentation is used to measure and improve care processes, especially those areas with recognized best practices such as documentation of risk assessment in prevention of patient falls during a hospital stay.

Documentation can consume a significant portion of nurses' time. Published estimates range from 20% to 50% of a nurse's shift, which can exceed the amount of time that nurses actually spend to provide direct patient care (Asaro and Boxerman 2008; Yee et al. 2012; Munyisia et al. 2013). Computerized documentation in an EHR has not improved documentation efficiency in general. In addition to efficiency concerns, nurses as users of EHRs have reported poor experiences such as downtimes and long computer system response times. The urgency to improve documentation efficiency in many health care organizations is justified by the increasing economic pressure on nurse productivity. These organizations are severely constrained by how much and how frequently the EHR they utilize can be modified. Baylor Scott & White Health embarked an improvement project for our current EHR in which cognitive engineering has contributed.

Despite numerous published reports about the time-consuming nature of nursing documentation, with or without an EHR, the understanding of the contributing factors to inefficiency and poor user experience is surprisingly inadequate. An important first step for us was to use methods and concepts that cognitive engineering can provide to uncover contributing factors, especially those with a high likelihood of being changed in a health care organization (e.g., limited ability to redesign the EHR, documentation policy and standards review, user support and education).

Methodological Considerations

We used three sources of information to guide our interview and observation efforts. The first source was hospital leaders, as sponsors of the improvement project and ultimate owners of patient outcomes. Hospital leaders took into consideration occurring safety concerns and other organizational contexts.

The second source was the frontline EHR support staff, who interact with nurses on a daily basis. In our organization, these EHR support staff are nurses themselves. The third source was informal consultations with frontline nurses. The types of units targeted were medical–surgical units (with a high patient/nurse ratio and are the dominant nursing care setting in our organization), ICUs (which have a large amount of patient care parameters to document), and neonatal ICUs (which treat a specialized population that have high demands of nutritional support). Two types of documentation activities were selected as the focus of study: documentation of a patient's assessment at the start of a shift and medication administration.

Contribution of Cognitive Engineering to Efficiency Improvements

To understand work as it was actually performed by frontline nurses, we conducted interviews with nurses at their workstations. Interview questions were designed to encourage nurses to perform a show-and-tell so that we could better understand specific areas in documentation that slowed them down, frustrated them, or had low value to patient care. Photographs and videos were taken with care to not capture any identifying information of patients or care providers. General assessment (e.g., how many times you had to stay after the shift to finish documentation) was combined with probing questions on specific situations (e.g., When did you start your documentation on shift assessment today, or how is today compared with the usual?). We specifically looked for prospective memory burdens (e.g., required assessment 30 min after pain medication), cognitive aids used (e.g., medication schedules), and team collaboration (e.g., physician orders placed many days in advance for a scheduled test or imaging study)

Walk-throughs with our nursing EHR support staff were useful to capture screens and to zero in on problematic documentation tasks. These EHR support staff were expert users and provided education to frontline nurses. They were able to simulate EHR use to allow screen capturing on specialized workstations, whereas a regular nurse on a typical clinical workstation usually was not able to do so. We asked the support staff to illustrate EHR areas that they received the most questions from staff or required detailed education efforts.

Macroergonomics: Communicating Insights with Key Stakeholders

Improvement takes resources. Sharing findings and achieving consensus on improvement strategies with a variety of stakeholders were judged to be an important step. Time–motion studies tended to provide quantitative data familiar to most stakeholders, even though sampling challenges tended to reduce the impact of such studies. Pictures, videos, and user stories were important to enhance the appreciation of hospital leaders for areas of improvement opportunities that were both general to all nurses and specific

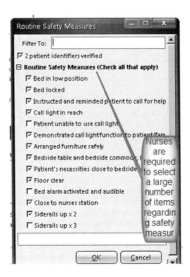

FIGURE 9.7
Example of screens to illustrate repetitive nature of documentation activities, as perceived by nurses. Because in theory, safety measures for each patient should be adjusted across time because of the risk profile changes, the EHR included a list that required nurses to document safety procedures for each patient. The first check box "2 patient identifiers verified" is a routine practice. Documentation does not improve compliance to safety protocols in general. Forcing nurses to routinely document such activities was perceived as low value.

to certain care areas. As an example, leaders were aware of staff complaints about duplicate data entry, yet they were told by EHR technical groups that these duplicate data entry issues had been removed from the EHR, and what remained was necessary documentation elements void of duplication. From the perspective of the frontline users, however, there seemed to be multiple places to document the same thing across several screens in the EHR. Although the multiple places were provided as convenience, many nurses believed that they had to enter the requested data at all the individual places available in the EHR. For example, to document a physical finding, users may have to navigate in repeated patterns multiple times, giving a feeling of duplication. Screen shots were useful for others to understand why nurses were continuing to complain about duplication (Figures 9.7 and 9.8; see Chapter 6 of this book for a related discussion).

Discussion

Shared understanding of documentation challenges enabled shared ownership of solutions by those who initiated changes in EHR, education efforts, and policy trade-offs. Discussions surrounding policy regarding documentation were initiated to clarify documentation requirements and to standardize documentation across the health system (e.g., so that physicians may find

Indication of anatomic sites 5 times

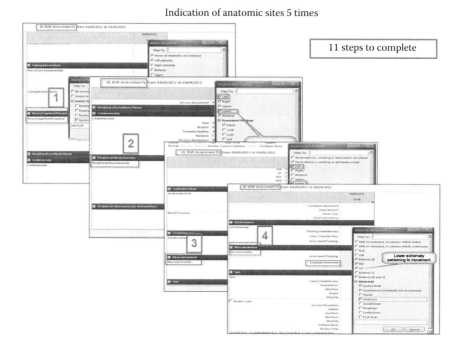

FIGURE 9.8

A sequence of screen shots to illustrate the nurses' perception of duplicate documentation. Documentation of the nursing care of a patient after total knee replacement procedure involved assessment of the same anatomic site (the surgical site) in different sections of the EHR such as the muscular–skeletal system, neuro-perceptual system, peripheral neuromuscular system, etc. A nurse had to navigate repeatedly to indicate the same anatomic site. A designer may not appreciate the perceived repetitive nature of the task. The sequence was used to enable hospital leadership to understand nurses' complaints.

information in consistent locations in the EHR). This case illustrates the range of problems and solutions that can be addressed by cognitive systems engineering. While it may be typical for cognitive engineers to focus on interface prototyping and interaction redesign, in this case, we supported the development of education and training strategies to target aspects of documentation that were confusing or difficult to learn. Like many other organizations, ours had a strong group of professional educators, and they benefited from insights gained from using cognitive engineering methods and concepts.

Case 4. Improving Infusion Medication Safety

A significant number of patients in hospitals receive medications through infusion pumps, which can be programmed to deliver medications over a period of time with a set rate. Internal estimates at our organization put that number at nearly 90%. Infusion pumps are used to deliver many high-risk, potent, and often life-sustaining medications to patients. Studies of adverse drug events have associated infusion medication with high error rates (Husch et al. 2005). However, infusion pumps with built-in safety mechanisms, such as dose error–reduction functions, otherwise known as "smart pumps," have not been shown to eliminate or even reduce adverse drug events (Rothschild et al. 2005). Dose error reduction functions are usually implemented through a "drug library," which is a list of medications with the possibility of including dose limits, default drug amounts, and default volumes. The drug library is a safety feature because dose errors outside limits may be flagged or stopped. A number of studies have illustrated defects in user interfaces that contributed to errors in programming infusion pumps (e.g., Graham et al. 2004). In one report, a usability evaluation did not result in the purchase of superior infusion pumps (Nemeth et al. 2009). Hospitals with smart pumps are seeking ways to improve safety, usually without the option of purchasing a new fleet of infusion pumps. Our human factors team was consulted to initiate improvement regarding the reduction of programming errors associated with high-risk medications as identified by the Institute of Safe Medication Practices.

Methodological Considerations

A field inspection with a nurse educator and a staff nurse was conducted to understand how infusion pumps were used by frontline nurses. The inspection revealed the time-consuming nature of pump programming. For example, to reach a life-support medication used frequently (vasopressin), a nurse had to scroll 11 times, with a response delay of 200 ms after each scroll. From the start to the end of programming, it would take an experienced nurse over 90 s to complete such a task. A subsequent analysis of the composition of the drug library provided further evidence that opportunities for improvement were available to reduce programming errors that could reach patients and to reduce difficulties in programming smart pumps.

The most important stakeholders were identified as the frontline nurses, who program smart pumps and "own" the responsibility of infusion medication safety. The human factors team identified a group of nurses and encouraged them to ask each other why the safety feature of programming the drug library that ensured that high-risk medications are administered at the appropriate clinical levels (e.g., volume and rate) was not used in programming the pumps. Examples were provided to the nurses in order to empower them to push beyond blaming individual nurse users. The barriers identified were surprisingly comprehensive and in line with findings in published reports. The nurses were then encouraged to present those findings to their peers as a list of "barriers" that impeded the use of the dose error–reduction system via the drug libraries (Figure 9.9).

Findings presented by ICU nurses

- Takes too long to program!

- Commonly used drugs are not easy to locate
  (e.g., Zosyn—piperacillin/tazobactam in medsurg, oncology and anti-infective CCAs)

- IVF/IVPB is not descriptive or intuitive for normal saline

- Unable to switch CCA libraries easily

- Not sure why most Abx/IVF are programmed

- Common doses are not included in CCA libraries

- Too many alerts (soft-limits)

- No organized ways to learn how to use the pumps exists

FIGURE 9.9
Sample slide used in discussions jointly led by frontline nurses and a human factors engineer. The findings were used to engage frontline providers to think broadly about barriers instead of an individual nurse's commitment to patient safety. IVF/IVPB, intravenous fluid and intravenous piggyback; CCA, clinical care area; ICU, intensive care unit.

The N = number of screen scrolls was calculated for the most frequently used medications to illustrate the need to reorganize the drug library (Figure 9.10). While this change was possible, the organization had not previously made the change due to a reliance on consensus and lack of understanding of human factors in listing many drug choices. The human factors team decided to invest time to develop an effective presentation of usability challenges such as the diagrammatic portraits of challenges in Figure 9.10. The leaders in the organization grasped the usability issue rapidly and overcame simplistic assumptions about the best ways to organize a drug library (such as alphabetical listing of all medications).

Macroergonomics: Engaging Multiple Stakeholders

This case illustrates the large number of stakeholders that can be involved in any specific problem and the need to deploy methods of macroergonomics and cognitive engineering that take a comprehensive integrative systems approach. For example, expectations for compliant use of the drug libraries during pump programming were set by administrative policies and guidelines. The skills necessary for programing infusion pumps, which were not intuitive, fell in the realm of professional nursing development. The service of pumps was provided by the biomedical engineering department. The pharmacy department held the responsibility for the drug library (with consultation and input from physicians and nurses). Routine reports on improvements to drug library usage, which were time consuming, were provided by the pump vendors and by an internal data analytics group.

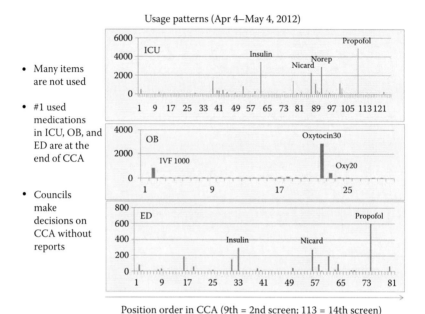

FIGURE 9.10

Sample slide used to share insight on organizational controlled variables that contributed to nurses' complaints that they took too long to program. The traditional assumption about the benefit of alphabetical listing of medications in the drug library was contrasted with information about frequency of drug use. BHCS, Baylor Health Care System; CCA, clinical care area; ED, emergency department; OB, obstetrics.

Engaging stakeholders in the analysis process provided a number of benefits. In addition to improving the pump programming process itself, frontline clinicians and managers learned about programming errors framed on the axiom that human errors are inevitable and are to be expected, and about the focus on minimizing patient harms as opposed to questioning their commitment to patient safety. A sample slide (Figure 9.11) was used for sharing the lessons learned in staff meetings. In addition to interviews and observations, cognitive engineering methods and concepts made essential contributions to safety improvements by providing visual representations to engage end users in designing reports to support frontline managers and other key leaders to rapidly improve drug library compliance, in utilizing reports to support drug library optimization, and in setting the systems view of improvement (Figure 9.11). Cognitive engineering played a role in developing the list of barriers and in the visual presentation of drug library usage logs. By bringing new concepts and solutions to discussions with key stakeholders, the human factors team was able to motivate systematic approaches to improve infusion safety.

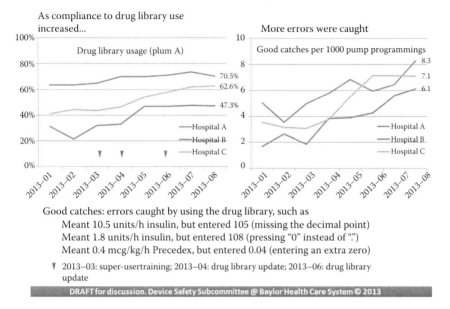

FIGURE 9.11

Example slide used to communicate to key stakeholders about success in improving "process" measures of safety and to encourage leaders to think more broadly about patient safety. The slide included three major improvement efforts and contrasted the improvement at three different hospitals. The slide was used to also illustrate how errors can occur not as a result of incompetence but as an expected part of human performance.

General Discussion

Health care organizations can benefit significantly from cognitive engineering concepts and methods, especially if they are applied broadly beyond simple device and interface design efforts and in conjunction with other applicable methods. Assessment of policies, education strategy development, and cultural change initiatives are among the top areas where cognitive engineering can lend methods and concepts about cognitive workload, expertise capturing, and performance support. In order to ensure a meaningful impact on the quality of patient care, strategies that engage multiple stakeholders are essential. Some of the methods based on cognitive engineering may be simplified and used by frontline clinicians, while others may be used to generate materials that help frontline clinicians and managers gain insight into cognitive work demands.

In the nursing workload project, cognitive engineering was used to frame the variables that contributed to a suboptimal work system. Improvements included optimizing the physical environment, the tools and technology used by nurses, the policies approved by the organization leadership, and the practices in complying with regulatory standards of patient care and documentation required by federal agencies. By framing workload as an interaction of technology, environment, people, and policies, we were able to move beyond leveraging more standard patient/nurse ratios, which are an important indicator but only one of several various mechanisms for improvement. As an example, artifacts used by nurses in their work can have an important impact on their workload (Xiao 2005). By looking at the paper forms used by nurses to facilitate their work, we were able to assess their workload as associated with prospective memory. As another example, we used the concept of interruptions to understand defects in our nurses' work system and developed a survey to assess the subjective impact of interruptions on their workflow as well as the common sources believed to be the cause of interruptions. The involvement of human factors specialists expanded the discussion of whether to hire more nurses to a systems level discussion centered on the concept of identifying bottlenecks and previously unknown sources of nursing workload.

In the improvement project in CV surgery, the perspectives held by different roles were shared during the cognitive walk-through we performed. Surgical teams were able to identify the specific needs of each role required for a safe completion of the CV surgical procedure. Observations across multiple surgeons' cases uncovered variations in terminology. Through the engagement of multiple stakeholders, the project team was able to secure the support from the manufacturer to prelabel CPB machine tubing at no additional cost to the system as an added safety measure. Additionally, a standardized terminology book was created and educated to CVOR staff which greatly reduced variability, confusion and cognitive demands placed upon staff that resulted from multiple surgeons utilizing different terminology.

In the improvement project on documentation efficiency, time–motion studies confirmed the complaints from nurses about documentation burdens, which comprise a large percentage of their time, in some cases up to 50% of a shift. Despite high hopes, EHRs have not yet resulted in a significant improvement to nursing documentation efficiency. While the literature contains numerous examples of how time-consuming documentation is for nurses, it lacks the fundamental understanding of contributing factors that extend documentation time for frontline nursing users. By going beyond the more traditional time–motion methodology, we were able to interview the users themselves in order to understand specific pain points to the timely completion of documentation. Staff provided feedback not only for specific usability concerns that slowed them down but also on how often they were forced to shift strategies to keep up with documentation requirements or even work in off-hours to catch up (i.e., before or after shifts). By better understanding the documentation challenges faced by our frontline staff, shared ownership in solutions was made possible: EHR redesign efforts, education efforts, and policy changes to reduce the burden of documentation and to provide more time for direct patient care activities.

In the case of infusion pump medication safety improvement, it was clear early on that multiple stakeholders would be needed to contribute to the solution. We found that it was productive to empower frontline users with rudimentary cognitive engineering concepts. This should serve as a reference for other health care organizations to use cognitive engineering methods to identify potential sources of barriers in using complex health technology. Hospital leaders and managers were shown usability challenges by visual portraits of the number of scrolls needed to program high-risk, time-critical medications. Cognitive engineering methods were useful to share insights and build a common ground for robust, effective improvement initiatives. Frontline nurses felt that their voices were heard when improvement in safety went beyond simply asking them to pay closer attention.

As demonstrated through our four case studies, engagement and macro-ergonomic approaches to solutions can help to overcome the inherent disadvantages faced by health care organizations to arrive at impactful, sustainable solutions. Building multidisciplinary teams that work in conjunction with the organization provides the mechanism for such solutions. Cognitive engineering provides a powerful arsenal of techniques that can help to uncover previously undiscovered safety threats, threats that an engaged frontline team, managers, and system leadership can target to improve the safety of the patients they care for and the quality of working life for their staff.

Acknowledgments

We thank Sarah Parker and Ann Bisantz for reviewing and providing feedback for this chapter. For the nursing workload case study, we thank frontline care providers and leaders (Rosemary Luguire, Don Kennerly, Jan Compton, Donna Montgomery, Claudia Wilder, Rita Fowler, Lynn Randolph, Toni Johnson-Akers, Caton Cadigan, Cortney Dalcour, Abigail Cartagena, Sandra Squires, Tiana Williams, Jason Jeffers, Ruby Relos, Rachel Jackson, Shameka Butler, Penny Quinn, Carol Crenshaw, Cindy Cassity, Richard Gilder, Cindy Wohlgemuth, and Chris Mata). For the operative room case study, we acknowledge the support and contributions of Myrna Dunn, Doug Robinson, Baron Hamman, Gonzalo Gonzalez-Stawinski, Mary Ann Guillien, and Nancy Vish. For the documentation case study, we thank Donna Montgomery, Chris McCarthy, Stephanie Bakker, Rincy Roy, Christie Burton, and Tamera Sutton. Finally, for the infusion safety case study, we extend our gratitude to the infusion safety task force members, especially Molly Hicks and Jason Trahan.

References

Adomat, R. and Hicks, C. (2003). Measuring nursing workload in intensive care: An observational study using closed circuit video cameras. *Journal of Advanced Nursing, 42*(4), 402–412.

Aiken, L.H., Clarke, S.P., Sloane, D.M., Sochalski, J., and Silber, J.H. (2002). Hospital nurse staffing and patient mortality, nurse burnout, and job dissatisfaction. *Journal of the American Medical Association, 288*(16), 1987–1993.

Asaro, P.V. and Boxerman S. (2008). Effects of computerized provider order entry and nursing documentation on workflow. *Academic Emergency Medicine 15*(10), 908–915.

Brady, A.M., Byrne, G., Horan, P., Griffiths, C., MacGregor, C., and Begley, C. (2007). Measuring the workload of community nurses in Ireland: A review of workload measurement systems. *Journal of Nursing Management, 15*(5), 481–489.

Carayon, P. (2007). *Handbook of Human Factors and Ergonomics in Patient Safety.* Mahwah, NJ: Lawrence Erlbaum Associates.

Carayon, P., Schoofs Hundt, A., Karsh, B.T., Gurses, A.P., Alvarado, C.J., Smith, M., and Flatley Brennan, P. (2006). Work system design for patient safety: The SEIPS model. *Quality and Safety in Health Care.* 2006 Dec; *15*(Suppl 1), i50–i58.

Carayon, P., and Gürses, A. P. (2005). A human factors engineering conceptual framework of nursing workload and patient safety in intensive care units. *Intensive and Critical Care Nursing, 21,* 284–301.

Graham, M.J., Kubose, T.K., Jordan, D., Zhang, J., Johnson, T.R., and Patel, V.L. (2004). Heuristic evaluation of infusion pumps: Implications for patient safety in Intensive Care Units. *International Journal of Medical Informatics.* 2004 Nov; *73*(11–12), 771–9.

Gurses, A.P., Martinez, E.A., Bauer, L., Kim, G., Lubomski, L.H., Marsteller, J.A., Pennathur, P.R., Goeschel, C., Pronovost, P.J., and Thompson, D. (2012). Using human factors engineering to improve patient safety in the cardiovascular operating room. *Work Stress, 41*(Suppl 1), 1801–1804.

Holden, R.J., Scanlon, M.C., Patel, N.R., Kaushal, R., Escoto, K.H., Brown, R.L., Alper, S.J., Arnold, J.M., Shalaby, T.M., Murkowski, K., and Karsh, B.T. (2011). A human factors framework and study of the effect of nursing workload on patient safety and employee quality of working life. *BMJ Quality and Safety, 20*(1), 15–24.

Hurst, K. (2005). Relationship between patient dependency, nursing workload, and quality. *International Journal of Nursing Studies, 42,* 75–84.

Husch, M., Sullivan, C., Rooney, D., Barnard, C., Fotis, M., and Clarke, J. (2005). Insights from the sharp end of intravenous medication errors: Implications for infusion pump technology. *Quality and Safety in Health Care, 14*(1), 80–86.

Joint Commission on Accreditation of Healthcare Organizations (2006). Tubing misconnections—a persistent and potentially deadly occurrence. Apr 3; *36,* 1–3.

Karsh, B.T., Holden, R.J., Alper, S.J., and Or, C.K. (2006). A human factors engineering paradigm for patient safety-designing to support the performance of the health care professional. *Quality and Safety in Healthcare, 15,* i59–i65.

Kazanjian, A., Green, C., Wong, J., and Reid, R. (2005). Effect of the hospital nursing environment on patient mortality: A systematic review. *Journal of Health Services Research and Policy, 10*(2), 111–117.

Kimehi-Woods, J. and Shultz, J.P. (2006). Using HFMEA to assess potential for patient harm from tubing misconnections. *Joint Commission Journal on Quality and Patient Safety, 32*(7), 373–381.

McGillis-Hall, L. (2003). Nursing intellectual capital: A theoretical approach for analyzing nursing productivity. *Nursing Economics, 21*(1), 14–19.

Morris, R., MacNeely, P., Scott, A., Treacy, P., & Hyde, A. (2007). Reconsidering the conceptualization of nursing workload: Literature review. *Journal of Advanced Nursing, 57*(5), 463–471.

Munyisia, E.N., Yu, P., and Hailey, D. (2013). Caregivers' time utilization before and after the introduction of an electronic nursing documentation system in a residential aged care facility. *Methods of Information in Medicine 52*(5), 403–410.

Neill, D. (2011). Nursing workload and the changing healthcare environment: A review of the literature. *Administrative Issues Journal: Education, Practice, and Research, 1*(2), 132–143.

Nemeth, C., Nunnally, M., Bitan, Y., Nunnally, S., and Cook, R.I. (2009). Between choice and chance: The role of human factors in acute care equipment decisions. *Journal of Patient Safety, 5*(2), 114–121.

Page, A. (2004). *Keeping Patient Safe: Transforming the Work Environment of Nurses.* Washington, DC: National Academies Press.

Pearson, A., Pallas, L.O., Thomson, D., Doucette, E., Tucker, D., Wiechula, R., Long, L., Porritt, K., and Jordan, Z. (2006). Systematic review of evidence on the impact of nursing workload and staffing on establishing healthy work environments. *International Journal of Evidenced-Based Healthcare, 4*(4), 337–384.

Poissant, L., Pereira, J., Tamblyn, R., and Kawasumi, Y. (2005). The impact of electronic health records on time efficiency of physicians and nurses: A systematic review. *Journal of the American Medical Informatics Association, 12*(5), 505–516.

Reason, J. (2000). Human error: Models and management. *British Medical Journal* (Clinical Research ed), *320,* 768–770.

Rothschild, J.M., Keohane, C.A., Cook, E.F., Orav, E.J., Burdick, E., Thompson, S., Hayes, J., and Bates D.W. (2005). A controlled trial of smart infusion pumps to improve medication safety in critically ill patients. *Critical Care Medicine, 33*(3), 533–540.

Simmons, D., Symes, L., Guenter, P., and Graves, K. (2011). Tubing misconnections: Normalization of deviance. *Nutrition in Clinical Practice, 26*(3), 286–293.

Wadhera, R.K., Parker, S.H., Burkhart, H.M., Greason, K.L., Neal, J.R., Levenick, K.M., Wiegmann, D.A., and Sundt III, T.M. (2010). Is the "sterile cockpit" concept applicable to cardiovascular surgery critical intervals or critical events? The impact of protocol-driven communication during cardiopulmonary bypass. *The Journal of Thoracic and Cardiovascular Surgery, 139*(2), 312–319.

Weydt, A.P. (2009). Defining, analyzing, and quantifying work complexity. *Creative Nursing, 15*(1), 7–13.

Wiegmann, D., Suther, T., Neal, J., Parker, S.H., and Sundt, T.M. (2009). A human factors analysis to cardiopulmonary bypass machines. *The Journal of Extra-Corporeal Technology, 41*(2), 57–63.

Xiao, Y. (2005). Artifacts and collaborative work in healthcare: Methodological, theoretical, and technological implications of the tangible. *Journal of Biomedical Informatics*, Feb; *38*(1), 26–33.

Xie, B. and Salvendy, G. (2000). Review and reappraisal of modeling and predicting mental workload in single-and multi-task environments. *Work Stress, 14,* 74–99.

Yee, T., Needleman, J., Pearson, M., Parkerton, P., Parkerton, M., and Wolstein, J. (2012). The influence of integrated electronic medical records and computerized nursing notes on nurses' time spent in documentation. *Computers Informatics Nursing, 30*(6), 287–292.

Index

Page numbers followed by f and t indicate figures and tables, respectively.

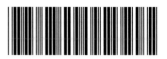